Hispanic Foods

ACS SYMPOSIUM SERIES **946**

Hispanic Foods

Chemistry and Flavor

Michael H. Tunick, Editor
Agricultural Research Service, U.S. Department of Agriculture

Elvira González de Mejia, Editor
University of Illinois

Sponsored by the
ACS Division of Agricultural and Food Chemistry, Inc.

American Chemical Society, Washington, DC

Library of Congress Cataloging-in-Publication Data

Hispanic foods : chemistry and flavor / Michael H. Tunick, editor, Elvira González de Mejia, editor.

　　p. cm.—(ACS symposium series ; 946)

"Developed from a symposium sponsored by the Division of Agricultural and Food Chemistry, Inc. at the 229th National Meeting of the American Chemical Society, San Diego, California, March 13–17, 2005."

Includes bibliographical references and index.

ISBN 13: 978-0-8412-3973-9 (alk. paper)

ISBN 10: 0-8412-3973-8 (alk. paper)

1. Food—Composition—Congresses. 2. Food—Analysis—Congresses. 3. Food—Sensory evaluation—Congresses. 4. Hispanic American consumers—Congresses.

I. Tunick, Michael. II. González de Majia, Elvira, 1950- III. American Chemical Society. Meeting (229th : 2005 : San Diego, Calif.) IV. American Chemical Society. Division of Agricultural and Food Chemistry, Inc.

TP372.5.H57　2006
664—dc22　　　　　　　　　　　　　　　　　　　　　2006048373

The paper used in this publication meets the minimum requirements of American National Standard for Information Sciences—Permanence of Paper for Printed Library Materials, ANSI Z39.48–1984.

PRINTED IN THE UNITED STATES OF AMERICA

Foreword

The ACS Symposium Series was first published in 1974 to provide a mechanism for publishing symposia quickly in book form. The purpose of the series is to publish timely, comprehensive books developed from ACS sponsored symposia based on current scientific research. Occasionally, books are developed from symposia sponsored by other organizations when the topic is of keen interest to the chemistry audience.

Before agreeing to publish a book, the proposed table of contents is reviewed for appropriate and comprehensive coverage and for interest to the audience. Some papers may be excluded to better focus the book; others may be added to provide comprehensiveness. When appropriate, overview or introductory chapters are added. Drafts of chapters are peer-reviewed prior to final acceptance or rejection, and manuscripts are prepared in camera-ready format.

As a rule, only original research papers and original review papers are included in the volumes. Verbatim reproductions of previously published papers are not accepted.

ACS Books Department

Contents

Indexes

Preface

This book is a result of the symposium *Chemistry and Flavor of Hispanic Foods* presented at the American Chemical Society (ACS) National Meeting held in San Diego, California on March 13–17, 2005. The symposium was sponsored by the ACS Division of Agricultural and Food Chemistry, Inc. with the goal of providing scientists and the food industry with the latest information on this expanding area. The Division is the leading scientific resource in the area of food chemistry and flavor.

Hispanics are the largest and fastest-growing minority in the United States and consumption of Hispanic-type food has been rapidly increasing. Some of the most popular foods, such as cheese, beans, and tea, have been subjected to little scientific investigation, which inhibits their use by food processors and in public food programs. This book covers these foods and others that are characteristic of Hispanic cuisine. The opening chapter details the Hispanic influence in restaurant menus, prepared foods, beverages, and flavors, and includes demographic and market data along with suggestions for food processors. Following chapters describe demographics of Hispanic snack food flavors (Chapter 2), chemistry of Hispanic dairy products, including cheeses, creams, yogurts, and desserts (Chapter 3) as well as a comparison of Mexican and European oregano (Chapter 4). Mexican peppers, including chipotle, are gaining in popularity in the United States, and two chapters on their flavor compounds are included (Chapters 5 and 6). The chemistry and biological activity of beans are described (Chapters 7 and 8), followed by a chapter on amaranth (Chapter 9), an ancient grain with nutraceutical properties. Lime is an important flavor in Hispanic cuisine, and the two major lime species are compared (Chapter 10). Two chapters (Chapters 11 and 12) on ethnic teas and their bioactive and aroma properties are included, and reveal that teas made from Ardisia plants have potential health benefits. The book closes with three chapters (Chapters 13–15) on chemistry, flavor, and volatile compounds in distilled beverages and margaritas.

We thank our authors for sharing their results with us in the symposium and in this book. We also thank the Division of Agricultural and Food Chemistry, Inc. for their financial support and for providing a forum for the symposium.

Michael H. Tunick
Dairy Processing and Products Research Unit
Eastern Regional Research Center
Agricultural Research Service
U.S. Department of Agriculture
600 East Mermaid Lane
Wyndmoor, PA 19038
mtunick@errc.ars.usda.gov

Elvira González de Mejia
Department of Food Science and Human Nutrition
University of Illinois
228 Edward R. Madigan Lab
1201 West Gregory Drive
Urbana, IL 61801
edemejia@uiuc.edu

Hispanic Foods

Chapter 1

Trends in Hispanic Foods

Abraham Wall[1] and Ana M. Calderón de la Barca[2]

[1]Kellogg Company México, S de RL de CV, Carr. Campo Militar km 1,
Querétaro, Qro., Mexico 76135
[2]Centro de Investigación en Alimentación y Desarrollo, AC, Carr. a la
Victoria km 0.6, Hermosillo, Son., Mexico 83000

Trends in Hispanic foods must be considered a bi-directional phenomenon. Hispanics within the U.S. are pushing for a spicier and fruity-flavored hybrid cuisine known as "Nuevo Latino." Many spices, herbs, flavors and homemade food ingredients from Latin American countries are increasingly being used by U.S. manufacturers. Conversely, if a food producer wants to target the Hispanic market, the option must be a product with a good combination of emotions and flavors mixed in a convenient food format. Product innovation for Latin America shall be regionalized using food ingredients recognized and accepted by the target market. The food service industry must learn from the "adopt and adapt" Hispanic philosophy because, if there is a multicultural factor that modifies America's food behavior, it is the Latino taste.

The "Hispanic" or "Nuevo Latino" cuisine represents an authentic culinary journey for Americans, while for Hispanics is a way to preserve their taste memory. Even in towns without a large Hispanic population, Americans are developing a taste for Hispanic food (1). With its excellent fruity-to-spicy flavor and many sensorial sensations conferred by their natural ingredients, Hispanic foods are nowadays one of the "big three" ethnic food trends in the U.S. (2).

A wide range of natural ingredients and dishes are gaining a place in supermarket shelves and restaurants: spices like oregano, pepper leaf and marjoram, teas like yerba mate and guarana, dairy-derived products like fresh cheese and the milk dessert "dulce de leche," and legumes and cereals like beans and amaranth, are just some examples. On the other hand, Latin Americans are changing their eating behavior due the nutritional transition that is affecting them (*3, 4*). However, introduction of new foods into their market should be carefully performed (*5*). Producers and restaurant owners must know that besides price, quality, and nutritional profile of the food offering, there are specific needs of the target consumer to be fulfilled, in order to design and launch successful products. This chapter is an attempt to contextualize the main drivers that make Hispanic foods so successful within the U.S. and across the whole American continent. Also, the basic rules that a food manufacturer must obey to launch a successful product targeted to the Hispanic market are discussed.

"We the U.S. Hispanics": Tracking Their Food Influence

Hispanics is the U.S.'s fastest-growing minority group, and so is its food popularity. Conservative data from the U.S. Census Bureau (*6*) states that 41 million residents in 2004 (around 14% of total population) were Hispanic, mainly of Mexican, Central and South American, and Puerto Rican origin. Also, Hispanic purchasing power is growing at triple the rate of the overall U.S. population. The market share controlled by Hispanic households rose from 5.2% in 1990 to 8% in 2004 (around $686 billion), mainly spent in food and other groceries (*7*), and geographically concentrated in New Mexico, Texas, California, Arizona, Nevada, Florida, Colorado, New York, New Jersey, and Illinois. The escalating population of Hispanics within the U.S., combined with the unlimited economic market potential they represent, deserves continuous market research for foodstuffs for this segment of the U.S. population.

"Hispanic" or "Latino" is not limited only to a country or primary language. It refers to all different cultures from Latin American countries, including the Caribbean ones. Although many other countries should be included (as the motherland Spain), the term is more a migration trend-derived word. Despite many differences in languages and culture habits, Hispanics share several characteristics to be used in marketing strategies. Hispanics are mostly young people of first (foreign-born) or second (U.S.-born) generation, with a strong identity (*8*). They spent 5½ hrs on-line weekly, like to travel for leisure (77%) or for visiting friends and family (43%), and are bohemian, passionate, and extremely family-oriented people (*2*). In terms of food and eating behavior, Hispanics are quite conservative, so they customize American foods into novel Hispanic dishes, defining the so-called "Nuevo Latino" cuisine. However, defining their cuisine is like jumping between different cultures and trying to harmonize them. Nevertheless, the restaurant industry terminology gives credit to South America (Brazil, Argentina, Ecuador, and Peru), Central America (Guatemala and Panama), Mexico, and Caribbean (Cuba and Puerto Rico), as exerting the greatest influence *(9)*.

Setting the Table: Restaurant Trends

Hispanics are constantly searching for their "Patria flavor." Ethnic foods are providing the impetus for the 6% growth in 2004 at independent casual-dining restaurants. According to the National Restaurant Association, Latino-themed restaurants are growing 3.5 times faster than any other restaurant group (even the pizza restaurants), and most mainstream restaurants are incorporating Latin fusions *(9)*. Hispanic restaurant firms like "Nacional 27" are gaining prestige among American restaurants, betting their success on the power of nostalgia of Latino immigrants and the potential appeal to Americans who want something other than burgers and fries. Also, many fast-food Latin American chains branched out to the U.S. after putting down strong enough roots at home. Led by Guatemala's "Pollo Campero SA", which opened its first U.S. outlet in 2002, many restaurant chains from other Latin America countries are expanding operations within the U.S. *(10)*. Some examples of the most widespread Hispanic foods and beverages invading the U.S. menus are shown in Table I.

Table I. Hispanic Foods and Beverages in U.S. Restaurant Menus

Country	*Food*	*Beverage*
Mexico	Tortilla soup	Tequila, Horchata
Panama	Carimañola	Chicha, Pipa
Guatemala	Fiambre	Cocoa smoothie
Puerto Rico	Asopao	Rum
Cuba	Yuca	Mojito
Brazil	Feijoada	Guarana, Caipirinha
Argentina	Churrasco	Yerba Mate tea
Chile	Choclos Cake	Pisco Sour
Peru	Ropa Vieja, Ceviche	Camu Camu tea
Colombia	Sancocho, Arepa	Aguardiente

Definitely, there are no boundaries for Hispanic foods. Americans are experiencing new and exciting food flavors after the introduction of lime-flavored seafood (ceviche), corn-based foods (arepa or tortilla soup), legume and grain-based soups (asopao, feijoada, and sancocho) and beefs (fiambre, churrasco, and "ropa vieja"), and tubercles (carimañola and yucca) and "ethnic veggies" (nopal) as side dishes (*10, 11*). Discovering a new Hispanic beverage is always a special pleasure, particularly if it is readily available in the U.S. market. Green coconut (pipa), nance (chicha), and rice (horchata) flavored beverages and popular alcoholic drinks from Chile (pisco), Colombia (aguardiente), Puerto Rico (rum), and Cuba (mojito), are also changing the America's palate (*11*). Whether "Latin" or Hispanic" foods means authentic dishes made with local ingredients or Latin ingredients to make a "twist" on several American dishes, this passionate cuisine is here to stay.

Catching the Flavor in Prepared Foods

Although it will take a longer time for "authentic Hispanic foods" to go from restaurant menus to the supermarket shelves, consumers are increasingly demanding authentic rather than quasi-Hispanic foods in grocery stores (*12*). Since taste is the most important factor to consumers for choosing a product, food producers have answered by adding new exotic ingredients and flavors to their current formulations (Table II). In that way, no other food segment, which includes Hispanic ingredients, has received more benefits than sauces and seasonings. Sold as spices alone (oregano, marjoram, pepper leaf, or yerba buena), blends (adobo, garlic mojo, sofrito, and mole), or sauces (jalapeño, chipotle), Hispanic spices and herbs are moving into a new era (*11, 12*). In addition, soup producers like Campbell's, Kraft, and Master Foods are directly importing Hispanic food ingredients from their country of origin and incorporating them in the final goods. Beans, tortilla, and cilantro are among the most cited Hispanic ingredients in new soup formulations (*13, 14*). Although the overall trend of these food segments seems to be falling (Table II), products based in ethnic ingredients (including Hispanics) are not.

The bakery and cereal industries are also suffering from this identity crisis (*8, 15*). From "conchas" (sweetrolls), "bollitos" (muffins), and "churros" (stick-shaped crispy pastry), American bakers have also kept in mind the Hispanic demographics. Hot flavors for pretzels (jalapeño), fillings for cakes and pastries (cream milk, cinnamon, and vanilla), and several fruity marmalades (guava, coconut, papaya, and guanabana) for donuts are just some examples. Ready-to-eat cereal manufacturers are also "following the cereal leader." The top three manufacturers in U.S. (Kellogg Co., Quaker Oats Co., and General Mills Co.) controlled 76% of the total cereal market in 2004, with its snack business the most promising (*16, 17*). The "whole grain wave" has came back with ancient cereal or pseudo-cereal grains like Mexican amaranth which is being incorporated in cereals, snack bars ("alegrias"), and even bread (*18*).

Table II. Hispanic Ingredients in Prepared Foods

	Country of Origin	2002	2003	2004
Soups		144	162	154
-Tortilla	Mexico	2	5	6
-Cilantro	Mexico	3	7	2
-Beans/green beans	Various	20	19	22
Sauces & seasonings		291	453	338
-Mole	Mexico	2	4	7
-Chipotle	Mexico	52	62	85
-Mojo	Cuba	4	3	3
-Sofrito	Central America	0	1	2
-Adobo	Several	4	1	3
Baked products & pastries		864	1,163	1,463
-Dulce de Leche	Several	5	10	9
Hot/Cold Cereals & Bars		107	153	177
-Amaranth	Mexico	10	9	7
Dairy Products		313	351	393

NOTE: Search performed as introductions between January and December and defined as "new formulation," "new product," or "new variety/range extension."

SOURCE: References 22, 23.

Dairy-derived product trends deserve special discussion. One-half of all dairy products new launches since 2002 are cheeses (*19*). According to the USDA (*20*), between October 2003 and October 2004, Hispanic cheeses jumped from 11.6 to 12.4% of the total cheese production in the U.S. The fresh and soft cheese is the most popular, particularly among Mexicans, who represent half of the total Hispanic population (*6*). It is used as a side dish or included in many popular Mexican foods ranging from "chiles rellenos" (cheese-filled green peppers) to "fritangas" (many corn-based popular foods, including "enchiladas" or "sopes"). However, due to the "home-made" nature of this product, its successful entry to the U.S. market is not reflected in new product launches (Table II). Another successful dairy-derived product is "dulce de leche" (milk dessert), which is widely recognized in many Latin American countries (*21*). Called "cajeta" (Mexico), "arequipe" (Venezuela and Colombia), or "manjar" (Chile and Argentina), this popular product has been included yearly in some new products since 2002 in candies, baked products, cereals, and ice creams.

It is noteworthy that Americans are more likely to try new food flavors than any other people in the world (24). In 2004, a higher proportion of ethnic flavor claims were made by U.S. manufacturers (12.6% of total SKU's) as compared to the 10.7% average from other regions (24). However, these spicy flavor trends clearly show that Americans are going to extremes. For example, the 134-year market leader in hot sauces, TABASCO®, has dominated the field with just one chili pepper based-product (25). Nowadays, this company offers 11 different products going from real "hot stuff" (Green TABASCO®) to mild hot and fruity sauces (chipotle and habanero TABASCO®). Table III shows several spices, herbs, and ethnic vegetables commonly included in prepared meals or condiments. While chipotle and jalapeño flavors travel from salty snacks (26) to ice creams (23), other spices like oregano, aji, and marjoram are more likely to be included just in prepared and seasoned meats (12).

Table III. Hispanic Spices, Herbs, and Veggies

		Spices		
Marjoram	Garlic	Sesame	Jalapeño	Habanero
Pepper Leaf	Aji	Oregano	Chipotle	
		Herbs		
Yerba Mate	Lapacho	Chamomile	Guarana	Marapuama
Damiana	Cilantro	Lemon grass	Linden mint	
		Vegetables		
	Nopal	Onion	Yucca	

NOTE: Search performed as introductions between January and December and defined as "new formulation," "new product," or "new variety/range extension."
SOURCE: References 22, 23.

While food manufacturers dream of a day when their products flow from supermarket shelves to restaurant tables and vice versa (like TABASCO® sauce), the "captured" Hispanic flavor probably is already positioned. Chicago-based Mintel's menu insights (27) identified in 2004 the Top Five Flavors in 550 restaurant menus, the sensory descriptors of "sweet" or "spicy": garlic (2,482 recalls), honey (652), barbecue (502), lemon (387), and herb (371). Excluding barbecue flavor, it can be said that all of these flavors have at least one Hispanic influence. For example, although garlic originally came from Western Europe, Mexico, Chile, and Peru contributed 27%, 3%, and 3% respectively to total U.S. garlic importation in 2003 (28). Taking care not to commit mistakes like "Taco Bell-ization" of Mexican tacos (1), food manufacturers can capitalize easily in successful debuts of flavors that are currently adopted by consumers. For instance, while Canada and Mexico fight for the "kidney bean exportation war" (*29*), "Tex-Mex" or "Mexican" cuisines capitalize on "enfrijoladas" (tortilla with fresh cheese covered with a blended-bean sauce) right now.

Beyond "Tequila" and "Mojito": Trends in Beverages

Traditionally, ready-to-drink (RTD) beverages were a source of refreshment, but the growing interest of U.S. consumers for beverages with additional benefits beyond thirst cessation has prompted manufacturers to look for new exciting flavors and ingredients with functional benefits (*11*). 2004's new beverage introductions revealed that companies are tying to match Hispanic flavors and ingredients in beverages to special consumer segments (*30*). Juices and concentrates remained the hottest trends in the category (Table IV), accounting for almost the half (total = 946) of all 2004 new beverage launches (*30*).

Table IV. Hispanic Ingredients in Beverages

	Country of Origin	2002	2003	2004
Beer & Cider		52	48	77
Flavored Alcoholic Drinks		21	19	27
Carbonated Soft Drinks		48	52	85
-Vanilla	Mexico	3	1	8
-Lime/Lemon	Mexico	6	13	11
Beverage Concentrates &		136	134	162
Mixes	Mexico	1	8	3
-Horchata	Mexico	5	10	5
-Jamaica	Mexico	2	7	3
-Tamarind	Mexico	17	18	20
RTD Coffee & Teas		36	41	53
-Yerba Mate	Argentina	5	12	10
-Hibiscus	Brazil	5	22	9
RTD Juices and juice drinks		144	197	293
Energy & Sports Drinks		68	65	101
-Guarana	Brazil	18	24	29
-Damiana	Brazil	4	4	1

NOTE: Introductions between January and December and defined as "new formulation," "new product," or "new variety/range extension."

SOURCE: References 22, 23.

Sports and energy drinks were the third growth driver, accounting for 11% of the total beverage category (*3*). This segment was the most likely to include ingredients like some holly plants from Brazil: guarana, lapacho, marapuama, and damiana extracts (*23, 30*). With more than $13.493 billion in sales in 2004, soft drinks occupied the fourth position (*30*). Vanilla, lemon, and lime flavors (Table V) add reasons of their own for popularity in soft drinks. Vanilla regular Coke and Lemon Diet Coke had a successful entry in 2002 with $208 and $139 million dollars in sales, respectively (*31, 32*). Since then, they have sustained their position in the U.S. market and across the border. In fact, Pepsi capitalized on Vanilla Coke's success, obtaining $125 million in sales while Lime Diet Coke (new launch) captured $119 million in 2004. New ready-to-drink (RTD) and concentrates "copy cats" from popular Latin American drinks are also growing like horchata, Jamaica, and tamarind. RTD coffee and herbal teas are gaining fame with claims like "Indian", "from the forest," or "mother nature." Yerba Mate and Hibiscus seem to be already positioned, while other herbs like ardisia and roselle will be more common as the "back to the basics" trend continues.

According to Mintel (*22*), the future will bring further fragmentation of the drinks category, where flavor is the main driver. Brands like Gatorade and Poweraid (sport drinks), Red Bull (energy drinks), Sprite, Coke, and Pepsi (soft drinks), and Lipton (teas) are targeting low-demographic U.S. populations with new "sensory sensations" (Table V). For many years, vanilla (*Vanilla planifolia*) extract was the most recognized flavor contribution from the Hispanic market. There are several types of vanilla, but the Mexican one can be easily identified by its typical, spicy, smooth, woody flavor and creamy consistency (*33*). Unfortunately, this "Totonacan vanilla" has been displaced in recent years by both French (Bourbon region) and Tahitian vanillas. To continue the success of citric flavors (lime and lemon), some other tropical flavors like guava, passion fruit, acai, and mango will jump into this trend in many flavor-mixed beverages including bottled water.

Table V. Top Fifteen Hispanic Flavors (*10, 23*)

Vanilla	Cinnamon	Acai	Guava	Coconut
Lime	Acerola	Cherimoya	Feijoa	Guanabana
Mango	Pineapple	Star Fruit	Papaya	Passion Fruit

Lastly, Hispanic alcoholic beverages are more than tequila: they represent a plethora of fruity-to-hot flavors from Patagonia to Tijuana. According to the America Economia Intelligence Group (*34*), among the Top 500 Latin American companies those from the beverage industry (most of them having beer brands) ranked higher than those dedicated to prepared foods (Table VI). "Natural" and "functional" trends have pushed the beer industry to include many new ingredients and flavors. From a technical perspective, these ingredients were used initially as masking agents to reduce beer bitterness to target a certain segment of consumers (women). However, this is not the case today: Anheuser-Busch Companies, Inc. recently launched two new beers, one lime and cactus flavored (9th Street Market Extra Special®) and another one Guarana-flavored (BE Beer®), while Steelback and Flying Dog Breweries launched a lime (Tango Beer®) and a chocolate flavored (Gonzo Beer®) beer. Meanwhile, there will be enough tequila brands (5 new launches per year) to prepare margaritas while the market for Colombian aguardiente, Cuban rum, and Chilean pisco will be gaining every day (23).

Table VI. Most Important Food Companies in Latin-America in 2004

Name	Rank	Country	Net Sales ($US millions)	Type
FEMSA	14	Mexico	8,426	Beverages
Bimbo	47	Mexico	4,623	Baked
AMBEV-CBB	52	Brazil	4,423	Beverages
Coca-Cola	55	Mexico	4,171	Beverages
Grupo Modelo	60	Mexico	4,019	Beverages
Coca-Cola	90	Brazil	2,788	Soft Drinks
Sadia	105	Brazil	2,403	Meat
Maseca-Gruma	116	Mexico	2,242	Cereal
Nestlé	123	Brazil	2,125	Cereal
Nestlé	127	Mexico	2,079	Cereal

SOURCE: References 34, 35

The Other Side of the Coin: New Foods for Latin Americans

Despite the notorious success of Hispanic foods in the U.S., launching new products for Latin Americans can be a real headache for food producers. When targeting a multicultural market, the country's economic indicators are less important than its population health trends, age, and lifestyles (5). For Latin Americans, purchasing food involves emotional, cultural, social, and even religious factors (1). The diversity of these factors between Latin American countries adds further complications when harmonizing food designs for Latin Americans as a whole. To put this in perspective, let us draft some conclusions from demographics and market data (Table VII). Despite their differences in total population, Peru and Brazil have the same human development index (HDI). However, Peru's agricultural-based economy is reflected in a lower food supply (460 Kcal/d/person) and food variety (20% less intake of non-starchy foods) than that observed in Brazil. The different eating behavior from Peruvians and Brazilians not only determined their differences in under nutrition prevalence (12% vs. 8.7%) but also in total grocery (0.86 vs. 26.3) and supermarket sales (0.17 vs. 16.8) which in turn affected differently the introduction of new food and beverages to their markets (5 vs. 194).

Table VII. Demographics and Market Environment in Latin America

	Parameters	MEX	CHL	PER	BRA	ARG	COL
1)	Population	103.4	15.5	28.0	180.0	38.0	41.0
2)	HDI	0.80	0.83	0.75	0.76	0.84	0.83
3)	TGR Sales	35.3	10.1	0.86	26.3	5.15	4.55
4)	Supermarket Sales	7.8	6.3	0.17	16.8	2.76	2.82
5)	Food Supply	3,160	2,850	2,550	3,010	3,070	2,580
6)	Non-Starchy foods	1,675	ND	1,173	1,987	1,996	1,522
7)	Under nutrition	5.2	0.6	3.4	15.6	0.6	5.7
8)	Total Food	109	1	3	145	132	38
9)	Total Drink	67	1	2	49	57	20
10)	Low Fat	3	0	0	8	7	1
11)	Low Carbohydrate	1	1	0	1		
12)	Trans Fat	0	0	0	0	0	0
13)	High Calcium	4	0	0	8	12	2
14)	High Iron	7	0	0	7	8	2
15)	High Fiber	1	0	0	5	4	1

NOTES: Mexico (MEX), Chile (CHL), Peru (PER), Brazil (BRA), Arg (ARG), Colombia (COL), Human development index (HDI), Total grocery retail (TGR), No data (ND). Values are expressed in millions (rows 1 and 7), $US billions (rows 3 and 4) and Kcal/d/person (rows 5 and 6). All data correspond to the 2002-2004 period.

SOURCE: References 6, 23, 29, 36, 37.

While Argentina follows the same trend as Brazil, Colombia is like Peru. Also, with different total population, Mexicans and Chileans shared almost the same HDI and prevalence of under nutrition, but Mexico's total retail sales was two times more that of Chile in 2002. Among Latin American countries, Chile, Cuba, and Costa Rica are the most nutrition and health conscious: the mortality rates of children under 5 years were below 10 in 2003, three to four times lower than in Mexico, Brazil, and El Salvador (38). After the U.S., Chile has the strictest regulations on nutritional labeling and health claims of the whole continent (*39*). So, it is not a surprise that new food product launches are highly controlled in this market as compared to Mexico. Definitely, if food producers want to be competitive in these "special" markets, they need to revise the needs of the target market first and then refresh their marketing strategies (*5*). To do so, there are five important considerations to take into account:

1. *Analyze local health trends and consumer awareness.* For Hispanics, health is highly correlated to food intake. Either by avoiding certain nutrients (e.g. fat) or by adding some others (e.g. vitamins), Hispanics look for health benefits in certain foods. For example, in many South American countries and Mexico (*40, 41*), iron deficiency and other micronutrient deficiencies have a high prevalence in children. Therefore, micronutrient-fortified products (like ready-to-eat cereals) will have a sustainable growth in LA markets (Table VI) for many years more.

2. *Speak the consumer's language: listen to what is wanted.* The high consumer awareness on chronic diseases such as cardiovascular diseases (*42*) contrasts with the low nutrition education of Hispanics. They will be more familiar with "foods that helps you to prevent diabetes" than some other "with low glycemic index." So, fiber-based foods will be more successful for preventing diabetes and obesity than their "Low-Carbohydrate" counterparts. Also, the introduction and sales of "light", "low," and "less of" brands growth will remain "slow".

3. *Add emotional and cultural factors.* New food product launches should have a "family", "country," and/or "religious"-related benefit, beside a convenient nutrient profile. The traditional Kellogg brand Corn Flakes® has gained consumer loyalty since its 2003 tagline "Corn Flakes in defense of those you love the most," offering this benefit to the whole family by its vitamin (antioxidant) profile.

4. *Surveillance of social and regulatory environment.* The success of a particular product in one LA country does not guarantee the same in another, as its introduction to the target market is restricted. One of the two beverages launched in Peru in 2004 (Table VI) was manufactured with coca leaf ("KDrink®" from Kokka Royal Food & Drink Co.), an ingredient highly restricted in mostly all LA countries.

5. *Customize flavors and food formats.* Kellogg's "Frosted Flakes" with arequipe flavor for Colombians, Slim Fast Bar Cinnamon/Banana Flavored for Brazil, and Goya Food's frozen loroco flower for Central American countries are just some examples of this rule.

Observing these basic rules, the only thing left is to "wait and see" if there is a successful return of the manufacturer's investment; Hispanics will respond with an enormous loyalty toward the manufacturer's food offering.

Summary

The Latino taste has strongly influenced the American way of eating. This is true not only for Hispanics but also for many Americans that look for new flavor and texture sensations. Food technology for these new foodstuffs is challenged continuously to deliver a wide range of flavors to resemble homemade Hispanic dishes. Latin American countries must be prepared for delivering these special ingredients to the global market since the possibilities for new combinations of Hispanic flavors are endless (1). However, food manufacturers must carefully pay attention to the Hispanics' needs. Wrong marketing strategies directed to this important segment of the U.S. population or Latin American market will result not only in less product sales but also in a complete divorce from one of the most loyal consumers in the world.

References

1. McGuire, N. Beyond the Taco. American Chemical Society Web Site, http://www.chemistry.org/portal/a/c/s/1/feature_ent.html?DOC=enthusiasts%5Cent_hispanicfood.html. Posted March 21, 2005.
2. Roberts, W.A. *Prepared Foods* **2005,** *174(5),* 13-21.
3. Mizón, C.C.; Atalah, E.S. *Rev. Chil. Nutr.* **2004,** *31(3),* 276-282.
4. Batista, F.M.; Rissin, A. *Cad. Saúde Pública, Rio de Janeiro* **2003,** *19,* 181-191.
5. Jensen, C.C. Today's Hispanic consumer. The Travel Industry Association of America Web Site, http://www.tia.org/uploads/casestudies/Hispanic%20White%20Paper.pdf. Posted April 27, 2005.
6. U.S. Census Bureau. "U.S. Interim Projections by Age, Sex, Race, and Hispanic Origin." U.S. Census Bureau Web Site, www. census.gov/ipc/www/usinterimproj/. Posted March 18, 2004.
7. Humphreys, J.M. *Georgia Business and Economic Conditions* **2004,** *64(3),* 11-12.
8. Bradley, H. *Baking & Snack* **2004,** *26(3),* 33-38.
9. Carinfa, M. Ingredient Challenges: Restaurant Trends on the National Menu. Prepared Foods Web site, http://www.preparedfoods.com/CDA/ArticleInformation/features/BNP_Features_Item/0,1231,159391,00.html. Posted September 1, 2005.

10. Bouza, T.; Sama, G. New Franchise Outposts Serve Changing Tastes. The Start Up Journal Web Site. http://www.startupjournal.com/ franchising/franchising/20030106-bouza.html. Posted January 6, 2003.
11. Sloan, A.E. *Food Technol.* **2000**, *54(9)*, 24-25.
12. Roberts, T. *Prepared Foods* **2005**, *174(3)*, 49-56.
13. Zind, T. *Prepared Foods* **2005**, *174(5)*, 77-79.
14. Roberts W.A. *Prepared Foods* **2005**, *174(3)*, 91-94.
15. Brinnehl, C. *Prepared Foods* **2005**, *174(3)*, 13-18.
16. Toops, D. *Food Processing* **2005**, *66(3)*, 37-40.
17. Brinnehl, C. *Prepared Foods* **2005**, *174(3)*, 31-36.
18. Archibald, A. *Prepared Foods* **2005**, *174(3)*, NS24-NS30.
19. Roberts, W.A. *Prepared Foods* **2005**, *174(5)*, 39-45.
20. National Agricultural Statistics Service. Dairy Products 2004 Summary. USDA, Washington, DC, 2005.
21. Smith, K.D.; Burrington, K.J. *UW Dairy Pipeline* **1999**, *11(2)*, 10.
22. Mintel's Global Product Data Base. Mintel Group Ltd Web Site www. gnpd.com/
23. Data monitor. Product Scan on Line. Marketing Intelligence Service Web Site www.productscan.com
24. Nosalik, D. *Prepared Foods* **2005**, *174(4)*, 13-19.
25. Gronlund, J. *Stagnito's New Prod Mag.* **2004**, *84.*
26. Katz, F. *Food Processing* **2005**, *66(3)*, 43-48.
27. Mintel's Menu Insights. Mintel Group Ltd. Web Site www. menuinsights.com/
28. Nation Master Com. Agriculture: Exports to USA Database Search. Rapid Intelligence Web Page http://www.nationmaster.com/graph-T/agr_exp_to_usa_gar&int=-1.
29. Nation Master Com. Agriculture: Exports to USA Database Search. Rapid Intelligence Web Page. http://www.nationmaster.com/graph-T/agr_exp_to_usa_kid_and_whi_bea
30. Roberts W.A. *Prepared Foods* **2005**, *174(3)*, 21-26.
31. Shelke, K. *Wellness Foods* **2005**, *7(1)*, 28-31.
32. Roberts W.A. *Prepared Foods* **2005**, *174(6)*, 12-17.
33. Rittman, A. *World Food Ingred.* **2005** (March), 32-34.
34. Samper, L.S. *Am. Economia* **2005** (Jul-Ago), 50-51.
35. Stock, G. *Am. Economia* **2005** (Jul-Ago), 44-45.
36. FAO. Food Security indicators. FAO Web Site. http://www.fao.org/es/ ess/faostat/foodsecurity. Posted 2005.
37. Business Monitor Internacional. Food & Drinks Reports. Business Monitor International Ltd. Web Site. http://www.businessmonitor.com /food/.

38. UNICEF. Estado Mundial de la Infancia 2005. Informe del Fondo de las Naciones Unidas para la Infancia. UNICEF Web Site. http://www.unicef.org/spanish/publications/files/SOWC_2005_(Spanish).pdf

39. MINSAL. Resolución exenta No. 556. Normas técnicas sobre directrices nutricionales que indica, para la declaración de propiedades saludables de los alimentos. Republica de Chile: Ministerio de Salud. http://www.minsal.cl/ici/nutricion/resolucion_exenta_556.pdf. Posted September 21, 2005.

40. Dávila, R.C.E.; Calderón, A.M.P.; Taipe, A.M.C.; Bernui, L.I.; Marlit Y.M.; Vanesa, R.D. *Rev. Peru Med. Exp. Salud Publica* **2004**, *21(2)*, 98-106.

41. Villalpando, S.; García-Guerra, A.; Ramírez, S.C.I.; Mejía R.F.; Matute, G.; Shamah-Levy, T.; Rivera, J.A. *Salud Pub. Mex.* **2003**, *45*, S508-S519.

42. Nugent, R. *Food Nutr. Bull.* **2004**, *25(2)*, 200-207.

Chapter 2

Hispanic Snack Flavors in the United States and Latin America

Patricia Rayas-Duarte[1], S. Corey Stone[1], Diana A. Freytez[1], Ana L. Romero-Baranzini[1], and Steven J. Mulvaney[2]

[1]Department of Biochemistry and Molecular Biology and Oklahoma Food and Agricultural Products Center, Oklahoma State University, Stillwater, OK 74078
[2]Department of Food Science, Cornell University, Ithaca, NY 14853

Snack foods are one of the fastest growing segments in the food market. Traditionally, new products in this category often involve new flavors. Therefore, it was decided to conduct a research survey in nine metropolitan areas throughout the U.S. and four Latin American countries in order to better understand snack food trends and flavor preferences. Categories surveyed were savory/salty, baked-goods/sweet, dairy, confectionary, and meat snacks. Highlights of the survey results are presented in this chapter. The survey consisted of twenty four questions and the participants were adults age 18 years and older and involved in at least half of the grocery shopping for the household. There were 1478 total participants, 54.7% from the U.S. and 45.3% from Mexico, Venezuela, Brazil, and Argentina. Salty (non-meat) snacks were by a large margin the preferred type of snacks in both the U.S. and Latin America over the categories of sweet/baked-goods, dairy, confectionary, and meat snacks. In the U.S., the preferred flavors of salty snacks were, in order of preference, cheese, pizza/Italian, and Hispanic/Tex-Mex. In Latin America, the preferred flavors of salty snacks were Hispanic/Tex-Mex, cheese, other flavors, and pizza/Italian.

Overall, the top two flavors of baked-goods/sweet snacks were chocolate and strawberry/other berries. The survey revealed opportunities to incorporate several tropical flavors with an established preference in Latin America that are not yet widely available in the U.S. snack food market. The development of healthier snacks (traditional, whole grains or fruit based) could benefit from the introduction of some of the flavors reported in this chapter.

Introduction

Snack foods are one of the fastest growing segments in the food market with estimated sales of more than $43 billion in the Americas in 2003 (*1, 2*). Together, Canada and the U.S. represent one of the largest snack foods markets with sales of more than $22 billion in 2002. Mexico and Central and South America are also important markets with total sales of $20.9 billion in 2001 (*1*). Hispanic flavors in snack foods continue to be among the most popular and provide an opportunity for introducing new flavors. The objective of this study was to investigate the flavor preferences of snack foods in the U.S. and Latin America. Food flavor preferences, and particularly snack flavors with a Hispanic flair, may provide direction for enhancing current food products, as well as for developing new healthier snacks containing functional ingredients designed for specific segments of the population.

Currently, Hispanics represent the largest minority population in the U. S. The U.S. Census Bureau projected that in 2005 that Hispanics will comprise 13.4% of the total population, followed closely by African Americans (12.6%), and a distant third by Asians and Pacific Islanders (4.3%) (*3*). By 2020, Hispanics will represent 21% of the total U.S. population while the African-Americans and Asians and Pacific Islanders are expected to be 12 and 6%, respectively. With a collective buying power of $560 billion, Hispanics outspend non-Hispanics across a variety of food categories. Hispanic families generally have lower-than-average incomes but spent about 20% more on groceries than non-Hispanic families ($430 and $320 per month on groceries, respectively) (*4*).

Snacks have become the fourth meal for a large portion of American consumers. This fact has been characterized by a re-definition of what constitutes a meal and what constitutes a snack food. In 1998, it was reported that about 75% of Americans consume snacks at least once a day (*5*).

Snack Food Trends in the Americas

Salted snacks are one of the biggest markets in the U.S., worth $23.5 billion in 2003 (*1*). The top three categories of chips were potato with $6.0 billion, tortilla/tostada with $4.5 billion, and meat snacks with $2.4 billion dollar sales. Other popular snacks are snack nuts/seeds/corn nuts, popcorn, pretzels, cheese snacks, corn snacks, and pork rinds. The U.S. exports and imports of salty snacks in 2003 were $337 million and $168 million, respectively (*6*). In the U.S., the salty snack industry has been growing with an average annual rate of 3.3%. Hispanic flavored snack foods have shown a growth trend; for example, sales of salsa in 2004 reached $22.4 million with an increase of 3.4% from 2003 (*6*). A variety of salsa products have reached grocery shelves in the Southern Plains region of the U.S. with the list of ingredient including more than the traditional basic tomato, onion and chili peppers. In addition, nouveau Latin cuisine can be easily found in upscale restaurants and supermarkets in medium and large metropolitan areas.

The Mexican snack food market is among the fastest growing import markets in the world and the U.S. is the largest exporter of snacks to Mexico. Snack imports from the U.S. in the year 2001 were $91.0 million, while Mexico's total snack imports from all countries totaled $92.0 million. Mexico produced snack foods worth $1.750 billion in 2001 (*7*). An estimated 22 million Mexicans had the purchase power and disposition to buy imported goods. Major Mexican snack food imports are popcorn, potato chips, peanuts, almonds, and other nuts including mixes. Mexico imported $20.6 million worth of potato chips in 2001 and $57.3 million worth of mixes, contributing the major share of the snack food imports (*7*).

The total sales of the Canadian snack market were $2.523 billion (Cdn $3.024 billion) in 2002 and $2.386 billion (Cdn $2.859 billion) in 2001 with an average growth of approximately 5.8% (*8*). Salty and baked snacks are more than 80% of the Canadian snack market. Baked snacks were on top of the list with $359 million in 2002, whereas the nuts and salty snacks were $301 million and $225 million sales, respectively. Potato chips and sticks were the preferred salty snacks among Canadians with total sales of $490 million. Corn chips were in the second position at $365 million. Popcorn, cheese snacks, and pretzels were $120 million, $85 million, $64 million sales, respectively (*8*).

Brazil is one of the largest markets in the world with a population of 168 million and a snack food market of $6.7 billion in 2000 (*2*). Processed food consumption is very high in Brazil. Both salty and sweet snacks form a traditional part of the Brazilian diet. The Brazilian snack market is evenly distributed into various categories, ncluding salty, biscuits and bakery, chocolate, confectionary, and dairy snacks. According to a Promar International survey report, total sales through chained retailers will be $75 billion by 2010 (*2*).

18

Venezuela is considered a growing market. The total Venezuelan salty snack consumption in 1997 was 15,000 metric tons per year with the major categories including pelleted and extruded products, corn and potato chips, popcorn, and processed nuts (9). Pelleted and extruded snacks and chips comprise close to 65% of the total snack consumption. Venezuela does not produce popcorn and depends totally on imports. Annual snack consumption and imports amount to 3,000 to 4,000 metric tons. The U.S. holds the major share of imports for Venezuela (9).

The above data reflect mainly sales and market size for snacks, but not flavor preferences in the major snack food categories. Therefore, it was of interest to compare the flavor preferences and frequency of consumption of different snack categories by American and Latin American consumers and analyze trends and effects of demographic variances in such preferences.

Materials and Methods

The snack food flavor preferences, frequency of consumption and demographics were studied using a 24-question survey that took about 3-4 min for the survey participants to complete. The interview instrument was prepared with input from Dr. Christine Johnson, Director of the Bureau for Social Research at Oklahoma State University. The interviews were conducted with adults age 18 years or older who participate in at least half of the grocery shopping for the household. Interviews were conducted from January through July of 2005. Nine metropolitan areas in the U.S. were surveyed, consisting of New York City, Chicago, Los Angeles, Miami, Dallas/Ft. Worth, San Diego, Denver, Atlanta, and Seattle. Four countries in Latin America were selected. They were Mexico, Venezuela, Brazil, and Argentina. Three cities were included in the surveys in Mexico (Mexico City, Cuernavaca, and Hermosillo) and Venezuela (Caracas, Puerto de la Cruz, and Puerto Cabello) and one each in Brazil (Florianopolis) and Argentina (Buenos Aires). The sample size (total number of respondents) was 1,478; 809 from the U.S. and 669 from Latin America. Statistical analysis of the effect of demographic variables in the snack preference was performed using chi-square analysis was used to determine significance of the effect of demographic variables on snack food preference. The results of the chi-square analysis for the U.S., Mexico, and Venezuela are discussed in this text. Results from Brazil and Argentina were inconclusive and therefore were not used for subsequent comparisons. This was due to smaller sample size (number of surveys) conducted in Brazil and Argentina compared to the ones from the U.S., Mexico, and Venezuela.

Results and Discussion

The preference patterns of snack types in the U.S. and Latin America are presented in Figure 1. Salty snacks were the first choice among the five snack types surveyed with preference ratings of 49% for the U.S. and 61% for Latin America. The baked-goods/sweet snack category was a distant second in preference. In the U.S., the preferences for baked-goods, dairy, confectionary and meat snacks were 22%, 14%, 10%, and 5%, respectively, while in Latin America the order was 15%, 11%, 12%, and 1%, respectively.

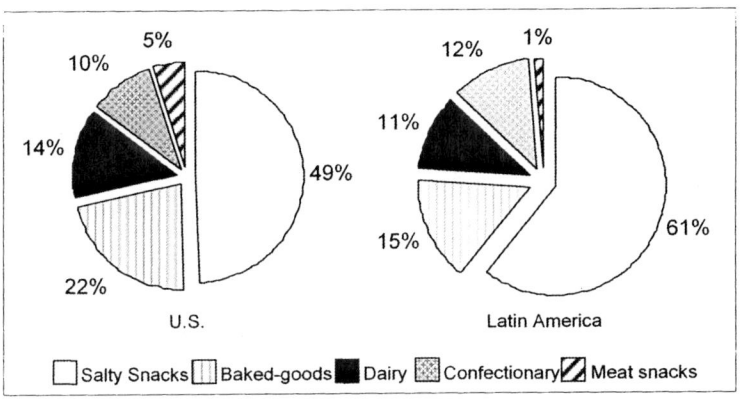

Figure 1. Preference rankings of major snack food categories in the U.S. and Latin America.

Salty Snacks

Flavor preference of salty snacks is shown below in Figure 2a. Preference in the U.S., in descending order was cheese, pizza/Italian, Hispanic/Tex-Mex, while in Latin America the order was Hispanic/Tex-Mex, cheese, other flavors, pizza/Italian. Cheese is clearly the number one salty flavor (39%) in the U.S., while in Latin America cheese tends to have similar preference (31%) to Hispanic/Tex-Mex flavors (33%).

The frequency of consumption of salty snacks in the U.S. is higher than that in Latin America (Fig. 2b). About 57% of Americans consume salty snacks two or more times per week compared to 30% of Latin Americans at that same frequency. Eighty percent of U.S. consumers eat snacks at least once per week while for Latin American consumers, this category is 13% lower at 67%. The results suggest an increase of about 5% of Americans consuming snacks at least once per week compared to 75% reported in 1998 (*5*). In 1999, 99.2% of U.S.

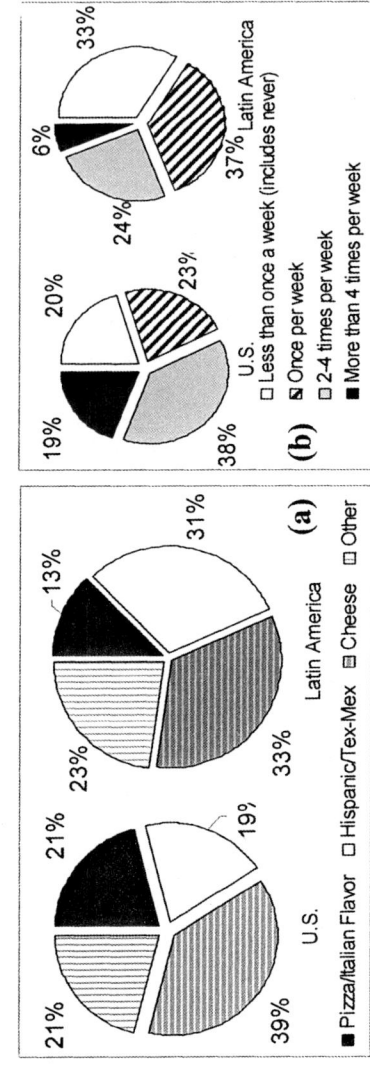

Figure 2. Flavor preference (a) and frequency of consumption (b) of salty snacks.

households purchased salty snacks, spending on average $76.39 yearly on 31.81 pounds (*10*).

Baked-Goods/Sweet Snacks

Overall, the top two preferred flavors of baked goods/sweet snacks in the U.S. and Latin America are chocolate and strawberry/other berry flavors (Fig. 3a). A distant third flavor of this category in the U.S. is apple (11%). In contrast, in Latin America the list of other preferred baked-goods/sweet snack flavors (43%) includes almond, vanilla, cinnamon, and butter. Interestingly, apple was not in the list. The question that addressed this flavor category in the survey had six choices plus an open ended question (other) to capture any other flavors not mentioned in the survey. Overall, the frequency of consumption for baked-goods/sweet snacks was higher in the U.S. compared to Latin America (Fig. 3b). About 49% of Americans consume baked-goods/sweet snacks more than two times per week compared to 31% of Latin Americans.

Dairy Snacks

The two top preferred dairy snack flavors in the U.S. were chocolate and vanilla with a similar rate of preference (36 and 34%, respectively) and a distant third was strawberry (15%, Fig. 4a). In contrast, in Latin America the top two flavors were strawberry and chocolate with similar rate of preference (33 and 32%, respectively) followed by vanilla (28%). Prominent in the list of "other favorite flavors" was mango. Overall, the frequency of consumption of dairy snacks in the U.S. and Latin America tends to be similar (Fig. 4b). The two frequencies receiving the highest scores were less than once per week and once per week with 47% and 24% for the U.S. and 39% and 29% for Latin America, respectively. These observations were surprising, since a higher frequency of consumption of dairy snacks was expected in the U.S.

It was interesting that the study revealed a relatively similar distribution of frequency of consumption of confectionary snacks in the U.S. and Latin America. About 56% of Americans and 66% of Latin Americans consume confectionary snacks once per week or more often (Fig. 5). About 44% of Americans consume confectionary snacks less than once a week (including never) as compared to 34% of Latin Americans. This could be explained by a cultural tradition of consuming caramel-type (with and without nuts) and tropical fruit confectionary in Latin America.

22

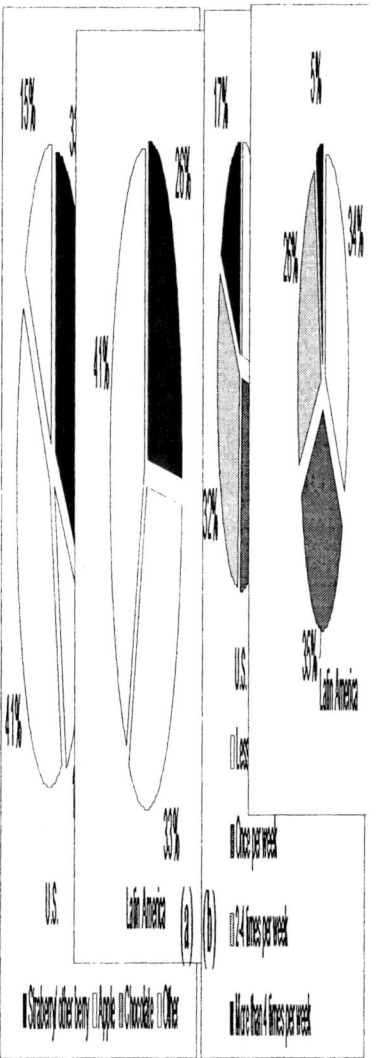

Figure 3. Flavor preference (a) and frequency of consumption (b) of baked-goods/sweet snacks.

23

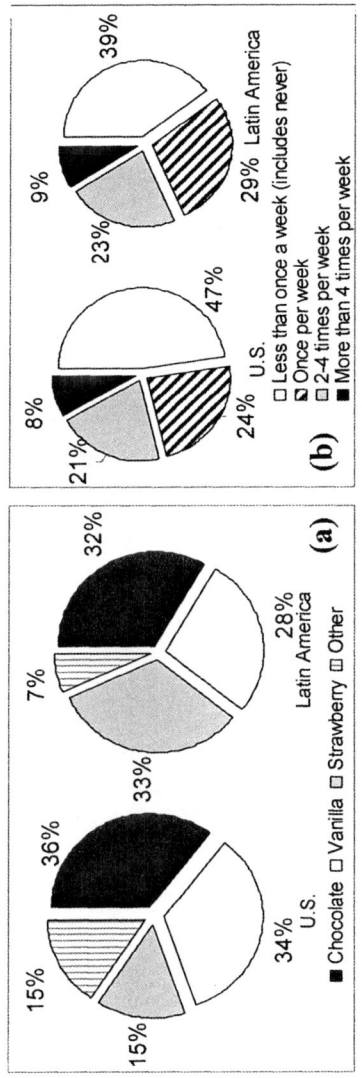

Figure 4. Flavor preference (a) and frequency of consumption (b) of dairy snacks.

24

Meat Snacks

Meat snacks represent a much smaller market in both the U.S. and Latin America. In the U.S., meat snacks are more common in the Southern Plains region. In both the U.S. and Latin America, meat snacks are perceived as being

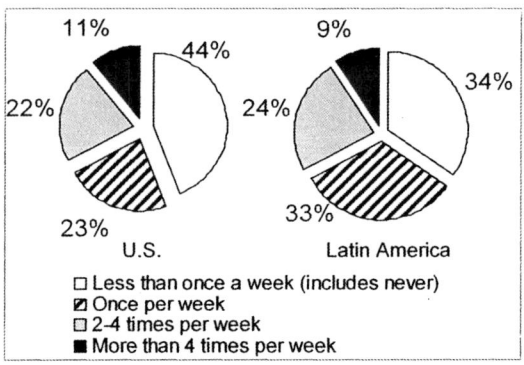

Figure 5. Frequency of consumption of confectionary snacks.

more popular with the young and adult male population. The preference of flavors in meat snacks was different in the U.S. and Latin America. The meat snack flavor preference in the U.S. was, in descending order, barbeque, plain, spicy, smoked. In Latin America the order was plain, spicy, barbeque, smoked (Fig. 6a). Meat snacks appeared to have similar frequency of consumption patterns in the two regions studied (Fig. 6b), with the majority of consumers (78 to 81%) reporting eating meat snacks less than once a week (including never).

Tropical Flavors

The top two preferred tropical fruit flavors found in this study were pineapple and mango with similar overall percentage rate in the U.S. and Latin America. Interestingly, banana (21%) was the third preferred flavor in the U. S. and coconut (12%) in Latin America. Among other tropical flavors mentioned in this survey were kiwi, passion fruit, lemon-lime and papaya. As expected, more consumers in Latin America (12% more than the U.S.) reported to have tried

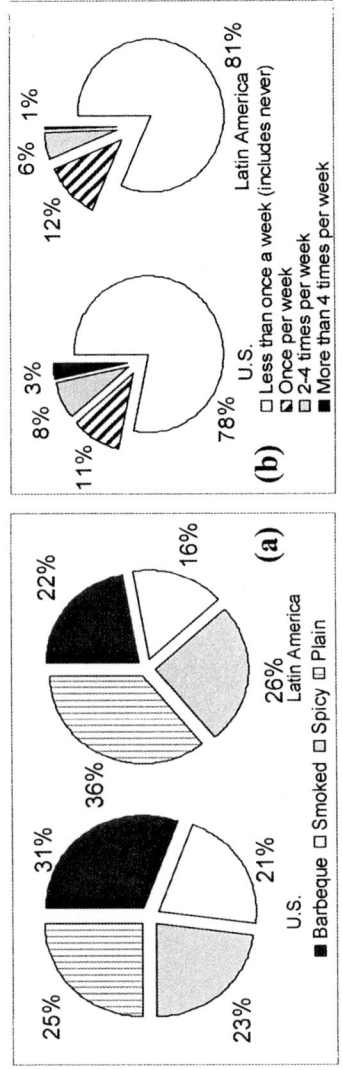

Figure 6. Flavor preference (a) and frequency of consumption (b) of meat snacks.

tropical flavor snacks. Perhaps the same percentage that chose "trying tropical flavors" could be a consumer base that might try other flavors as well. This may represent an opportunity to explore tropical flavors for new products. These tropical flavors might also set new snack products apart if they contain other perceived (or better yet, actual) nutritional advantages due to inclusion of functional and/or natural ingredients (i.e., nutraceuticals or pre/probiotics). These opportunities would be most likely explored by small/medium size companies with more versatility and innovative outlook and focus on niche markets.

Level of Spiciness

In reference to the level of spiciness in salty snacks, it was interesting to note a trend towards "extra hot" spiciness in the U.S. (10% higher than Latin America), while the "medium" spiciness level was more popular with Latin Americans (17% higher than the U.S.). But these observations still require clarification regarding the specific sensory perception of spiciness levels in each country and individuals, since it can vary widely. However, the overall trend suggests that Americans continue to explore spicy flavored snacks.

Effect of Demographic Variables on Snack Preferences

In the following section, a summary of the effect of demographic information (i.e., age, marital status, etc., independent variables) on snack category and flavor preferences (dependent variables) using the chi-square analysis method is presented (Tables I-IV). The analyses were done using surveys from the U.S., Mexico, and Venezuela. The sample sizes from the surveys from Brazil and Argentina were not large enough to provide statistical significance when using the chi-square analysis.

Table I suggests that the independent variables (demographic factors) affected a greater number of dependent variables (frequency of consumption, flavor of snacks, etc.) in the U.S. sample as compared to the samples from Mexico and Venezuela (Tables II and III, respectively). Race and salary in Mexico and race in Venezuela did not affect the dependent variables. Summarizing by column, in the U.S. age affected a greater number of dependent variables (13), followed by race (11), and marital status, Hispanic heritage, and gender (10 each) and number of children living at home (8) (Table I). Similar to the U.S., in Mexico age affected more dependent variables (14) related to the frequency and flavor of snacks. But in contrast to the U.S., in Mexico three other demographic variables affect the selection of snacks and flavors consumed.

Table I. Effect Of Independent Variables (Demographics) on Preference for Snack Category and Flavor (Dependent Variables) in the United States: Chi-Square Analysis at 90, 95, and 99% Confidence

Independent[a] / Dependent[a]	Age	Marital Status	Children[b]	Hispanic Heritage	Race	Salary	Gender	MA[c]
Type of Snack Food					*		*	
Consumption Frequency								
Salty Snacks	***	***	*	***	**	**		
Baked Snacks			**	**				
Dairy Snacks	*		*	**	***		*	
Confectionary Snacks	**		*					
Meat Snacks	**	***	***	***	***	***	***	
Flavor preference								
Salty Snacks	**	*		***	***	*	**	
Baked Snacks	***				**		**	
Dairy Snacks	*			***	***		*	**
Meat Snacks	***	***	*	***	***	**	**	
Favorite Tropical Flavor	***	**		***	***		**	**
Tried New Tropical[d]	***	***	**					
Mango		***						
Pineapple						*	**	
Passion Fruit	***	**						
Lemon/Lime				*	*			
Tried Other Flavor		**						**
Level of Spiciness	***	**	***	***	***	**	***	*
Tried New Snacks	*							

[a]Dependent and independent variables. [b]Number of children living at home. [c]MA = Metropolitan area. [d]Have tried new tropical flavored snacks. * P < 0.10, ** P < 0.05, and *** P < 0.01.

These were marital status, number of children living at home, and Hispanic heritage (5 variables each), with the third demographic, gender, affecting 3 variables (Table II). These observations contrast with Venezuela, where gender appears to have more effect on the selection of snacks and flavors (9), followed by number of children living at home and salary (7 each), and age (5) (Table III). The country of residence significantly affects most of the dependent variables except trying new snack flavors (Table IV). The data suggest that there is no significant difference in the decision of trying new flavors based on country of residence.

Table II. Effect of Independent Variables (Demographics) on Preference for Snack Category and Flavor (Dependent Variables) in Mexico: Chi-Square Analysis at 90, 95, and 99% Confidence

Independent[a] / Dependent[a]	Age	Marital Status	Children[b]	Hispanic Heritage	Gender
Type of Snack Food	**	***			
Consumption Frequency					
Salty Snacks	***		*		
Dairy Snacks	*			**	
Meat Snacks				**	*
Flavor preference					
Salty Snacks	***				
Baked Snacks	*				
Dairy Snacks			*		
Meat Snacks	***				
Favorite Tropical Flavor	***		*	***	**
Mango			*	***	
Coconut	**				*
Pineapple			**		
Passion Fruit	***	**			
Lemon/Lime	***				
Papaya	**	**			
Tried Other Flavors					
Level of Spiciness	***	**			
Tried New Snacks	*	**		***	
Tried New Flavors	**				

[a]Dependent and independent variables. [b]Number of children living at home. * $P < 0.10$, ** $P < 0.05$, and *** $P < 0.01$.

Conclusions

As expected, salty snacks are by far the main snack type in both the U.S. and Latin America with 49 and 61% preference rate, respectively. But this survey demonstrates just how far ahead in preference salty snacks really are. Cheese is the top flavor of salty snacks for Americans and it appears to be equally preferred

Table III. Effect of Independent Variables (Demographics) on Preference for Snack Category and Flavor (Dependent Variables) in Venezuela: Chi-Square Analysis at 90, 95, and 99% Confidence

Independent[a] / Dependent[a]	Age	Marital Status	Children[b]	Hispanic Heritage	Salary	Gender
Type of Snack Food	***		***		**	
Consumption Frequency						
Baked Snacks			**			
Dairy Snacks			**			*
Confectionary Snacks	**		*		***	
Meat Snacks				*		
Flavor preference						
Baked Snacks					**	
Dairy Snacks		**	**			*
Meat Snacks	*				***	**
Favorite Tropical Flavor						**
Mango	*					**
Coconut					*	
Kiwi						*
Passion Fruit			*		**	
Banana						**
Papaya	**	**				
Tried Other Flavor		*				**
Tried New Snacks						**
Tried New Flavors			*		**	

[a]Dependent and independent variables. [b]Number of children living at home. * $P < 0.10$, ** $P < 0.05$, and *** $P < 0.01$.

to Hispanic flavors for Latin Americans. The projected growth of the Hispanic population in the U.S. suggests that we will see more Hispanic flavored snacks in the U.S. as well. Chocolate and strawberry and other berries are the top flavors of baked-goods and dairy snacks in the U.S. But banana was the third preferred flavor in the U. S. (21%), while coconut (12%) was third in Latin America. Pineapple and mango are the top tropical fruit flavors based on results of this survey. However, other flavors showing good rates of preference were coconut and banana, followed by kiwi, passion fruit, lemon lime, and papaya. The preferences in the fruit flavors could be very useful in developing fruit-based functional foods for different geographic market segments, while it remains to be seen if the flavor preferences for traditional salty snacks will also apply to new whole-grain salty snacks, or products combining whole grain and fruit.

Table IV. Effect of Country of Residence (Independent Variable) on Snack Food Preferences (Dependent Variables): Chi-Square Analaysis at 90, 95 and 99% Confidence

Dependent variables / *Independent variable*	*Country of Residence*
Type of Snack Food	***
Consumption Frequency	
Salty Snacks	***
Baked Snacks	**
Dairy Snacks	***
Confectionary Snacks	***
Meat Snacks	***
Flavor preference	
Salty Snacks	***
Baked Snacks	***
Dairy Snacks	***
Meat Snacks	***
Favorite Tropical Flavor	***
Have Tried Tropical Flavored Snacks	***
Mango	***
Pineapple	***
Coconut	***
Kiwi	***
Passion Fruit	***
Lemon/ Lime	***
Banana	***
Tried Other Flavor	***
Level of Spiciness	***
Tried New Snacks	***
Tried New Flavors	

* P < 0.10, ** P < 0.05, and *** P < 0.01.

Acknowledgments

Research was supported by the Oklahoma Agricultural Experiment Station and the Food and Agricultural Products Research and Technology Center, Oklahoma State University. The technical assistance of Alejandra Penaloza-Encarnacion, Pauwei Yeap and Malla R. Devarapalli is greatly appreciated.

References

1. Snack Food Industry Trends; *Snack Food & Wholesale Bakery*, **2005**; http://www.snackandbakery.com/content.php?s=SF/2004/08&p=14; Access date: 6/7/2005
2. The Future of Snack Markets in Latin America; *Promar International*, **2001**; http://www.promarinternational.com/.../BrandedFoodBeverages/LA% 20Snacks%20Management%20summary.pdf
3. United States-Profile of General Demographic Characteristics: 2000, *U.S. Census Bureau*, **2000**; http://censtats.census.gov/data/US/01000.pdf
4. Hernandez-Medina, E. *Hispanic PRWire*, **2003**; http://www.hispanicprwire. com/news_in.php?id=470&cha=4&PHPSESSID=a8b8a6d8b3bb12c8a184e 4a10e624c25 .Access date: 02/15/2005
5. *American Dietetic Association*, **1998**; *Snack Attacks are Okay*, Brochure Number: 7106; The American Dietetic Association. Also available at http://www.clemson.edu/nutriweb/library_detail.php?libraryID=919. Access date: 08/22/2005.
6. Snapshot U.S. Market Trends – Salted Snacks, April **2005**; *USDA, Foreign Agricultural Service, Processed Products Division*; www.fas.usda.gov/ agx/ISMG/SnapshotSaltedSnacks0405.pdf
7. Mexico Product Brief- Snacks Foods Market Brief, **2003**; GAIN Report Number: MA3310; *USDA Foreign Agricultural Service GAIN Report,Global Agriculture Information Network*; http://www.fas.usda.gov/ gainfiles/200306/145985303.pdf
8. Canada- Market Development Reports, Snack Food Market in Canada, **2003**; GAIN Report Number:CA3006; *USDA Foreign Agricultural Service GAIN Report, Global Agriculture Information Network*; http://www. fas.usda.gov/gainfiles/200301/145785163.pdf
9. Market Brief: Snack Foods (Dry, Salty)-Venezuela **1997**; Report Code VE9724v; *USDA Foreign Agricultural Service; FASonline*; p. 1-13. http://www.fas.usda.gov/scripts/attacherep/display_gedes_report.asp?Rep_ D=1... Access date: 02/21/2005
10. Kuchler, F.; Tegene, A.; Harris, J. M. *Rev. Ag. Econ.* **2005**, *27*, 4-20.

Chapter 3

Hispanic Dairy Products

Michael H. Tunick

Eastern Regional Research Center, Dairy Processing and Products Research Unit, Agricultural Research Service, U.S. Department of Agriculture, 600 East Mermaid Lane, Wyndmoor, PA 19038

Hispanic-style cheeses and other dairy products are increasing in popularity in the U.S., prompting research into the chemical basis for their characteristics. Variations in Hispanic-style cheeses arise from differences in their processing parameters, which allow the properties of the cheese to range from hard and strongly flavored (Cotija), through semi-hard and meltable (Asadero, Oaxaca, and Mennonita), to soft and crumbly (Panela, Queso Blanco, Queso Fresco, and Requesón). Other dairy products that are prominent in Hispanic cuisine include creams (Crema Mexicana, Crema Agria, Natas), yogurts (cup, drinkable, and Jocoque), and desserts (Dulce de Leche, ice cream, Tres Leches). The diverse attributes of these products can be defined by the science underlying them.

Hispanics comprise 13.3% of the U.S. population, with two-thirds identifying themselves as Mexican (*1*). U.S. food companies are now targeting Hispanics (*2*) as well as non-Hispanics who are interested in trying Latin American food (*3*). Dairy products, unknown prior to Spanish colonization, are widely consumed in Mexico (*4*) and the rest of Latin America. The market in the U.S. for Hispanic cheeses (*5*), yogurts (*6*), and desserts (*6*) is growing, but the available literature on their chemistry and flavor is scant. A 1990 FAO report on dairy products in developing countries covered cheese, yogurt, and some desserts (*7*). More recently, Van Hekken and Farkye reviewed major Hispanic cheese varieties (*8*). A recent study revealed that over 85% of Mexicans and Mexican-Americans consumed cheese (*4*), but most of the literature on Hispanic cheeses has dealt with bacterial contamination rather than the cheeses themselves. The consumption of other Hispanic-style dairy foods has also risen. Creams are an important part of Hispanic cuisine and their consumption has increased in the U.S. (*9*), but no papers covering these products have been published. Cup yogurt, drinkable yogurt, and Jocoque are products that are also popular among Hispanics. Dulce de Leche, Tres Leches, and ice cream are examples of dairy-based Hispanic desserts, but research on these has also been limited. The purpose of this chapter is to describe representative Hispanic dairy products, emphasizing the information published about their chemistry.

Cheeses

Hispanic cheeses can be classified into three categories, hard, semi-hard, and soft. The hard varieties are grating cheeses that tend to be strongly flavored, the semi-hard cheeses are less flavorful and usually made to be meltable, and the soft cheeses are often bland and crumbly. The flavor and the textural and functional properties of cheese rely on levels of moisture, casein, fat, salt, pH, and other factors. When the moisture content is low, the casein network is too dense to melt or stretch. No melting or stretching takes place in cheese until the pH drops below 6.0. At pH 4.8-5.8, H^+ replaces $CaPO_4$, casein micelles fuse into a continuous network, and (especially at pH 5.0-5.4) cheese can melt and stretch. When the pH is lowered to 4.6, casein molecules aggregate again, the ability to stretch is lost, and meltability is reduced (*10*).

The pH of milk in cheesemaking is sometimes reduced by the addition of food grade acid, but is traditionally lowered by the lactic acid resulting from bacterial metabolism of lactose. Many Hispanic cheeses are prepared with raw milk, which contains indigenous lactic acid bacteria. Raw milk cheese is not permitted in the U.S., so companies there use pasteurized milk, as do many of the larger cheese plants in Latin America. Since pasteurization inactivates microorganisms, starter cultures containing suitable bacterial strains must be added. Coagulation is performed with rennet, which traditionally contains chymosin enzyme derived from calf stomach but nowadays is likely to be derived from bacterial sources such as *Rhizomucor miehei*. The enzyme cleaves κ-casein, destabilizing the casein micelles and allowing them to bond to each other and form a curd.

Hispanic cheese names are not standardized in the U.S. (*11*) or Latin America (*12*), leading to generic and trademarked brands that may not correspond exactly to the varieties below. For instance, Queso Añejo is used to describe both aged Cotija and aged Queso Blanco.

Hard Cheeses

The Hispanic cheeses in this category are represented by Cotija and its variations. The composition of Cotija is shown in Table I. Comparable to Parmesan, it is made with cow milk (or often goat milk in Latin America), thermophilic starter cultures, lipase enzymes for lipid breakdown, and $CaCl_2$ to assist in coagulation. Thermophilic starter cultures are most active above room temperature, with *Streptococcus thermophilus* and *Lactobacillus helveticus* being common examples. Cotija is cooked at 65-70°C for 30 min and coagulated by rennet at 32-35°C in 40 min (*13*). The curd is milled and heavily salted, imparting a very salty flavor. The high manufacturing temperature and salt content lower the moisture level and makes Cotija a non-melting cheese with a strong casein matrix that results in a hard, brittle texture suitable for grating.

Table I. Composition of Cotija (*8, 14, 15*)

	Cotija
Moisture (%)	35-42
Fat (%)	23-30
Salt (%)	4.7-5.4
Protein (%)	28-31
pH	4.7-5.5

Semi-Hard Cheeses

Oaxaca, Asadero, and Mennonita are examples of semi-hard Hispanic cheeses; their composition is shown in Table II. All are coagulated by rennet from whole milk after the addition of mesophilic starters (bacteria such as *Lactococcus lactis*, which are most active at room temperature), and all melt and slice well (*8*). Cooking of Oaxaca and Asadero curd takes place before draining all of the whey. Oaxaca is prepared and kneaded like Mozzarella, pulled into a strand, dry salted, and rolled into a ball (*8, 16*). Asadero, "fit for roasting," is pre-acidified by addition of dilute acid or acid whey. Traditionally, half of the milk is soured overnight and added to fresh milk (*17*) before rennet coagulation. This cheese is stretched, molded into a log or sphere, and sold fresh (*8*). The stretching step aligns the casein matrix, allowing the cheese to string when heated. In the U.S., Asadero is usually a processed blend to enhance melting. Oaxaca has a sweet milk flavor (8) and Asadero is slightly tangy and buttery (12).

Table II. Composition of semi-hard, meltable cheeses (8, 14, 15)

	Asadero	Oaxaca	Mennonita
Moisture (%)	41-49	40-46	41-45
Fat (%)	18-30	23-25	21-30
Salt (%)	0.8-1.9	1.4-1.8	1.4-2.3
Protein (%)	21-30	24-27	22-26
pH	5.3-5.6	5.0-5.5	5.0-5.3

Mennonita is manufactured in a manner similar to Cheddar, but contains more moisture. The curd is coagulated with rennet in 15-40 min, cut, and cooked at 37-45°C for up to 30 min. After whey drainage, the curd is cheddared – slabs of curd are stacked to promote additional whey removal (18). Mennonita has a shelf life of only 1-2 mo, and the product is generally eaten sooner. Both pasteurized and raw milk are used to make Mennonita in Mexico, and the resulting differences in the bacterial counts in the cheeses lead to differences in protein breakdown and rheological properties. Pasteurized milk Mennonita does not vary with season and undergoes less proteolysis than raw milk Mennonita, leading to differences in appearance and texture. Raw milk cheeses exhibit gas holes resulting from enzymatic activity, but these openings are virtually absent in pasteurized milk cheeses, which contain far fewer viable microorganisms. Raw milk cheese has seasonal effects, with winter cheese undergoing less casein proteolysis (19). Cheese made from raw milk in the summer is harder and springier than in other seasons because of its elevated protein content (19). A version of Mennonita is made in the U.S. from pasteurized milk, and is usually called Queso Quesadilla and Chihuahua™.

Soft Cheeses

Panela, Queso Fresco, Queso Blanco, and Requesón are soft Hispanic cheeses that can be sliced and crumbled. They traditionally have relatively high pH values and are therefore not meltable. Table III lists the compositions of soft Hispanic cheeses.

Panela is also called Queso de Canasta, "basket cheese" (20). It is coagulated by rennet set from pre-acidified whole cow milk (or, in Mexico, a combination of sheep and goat milk) with some acid added, and is cooked at 30-

Table III. Composition of soft, crumbly cheeses (*8, 14, 15*)

	Panela	*Queso Fresco*	*Queso Blanco*	*Requesón*
Moisture (%)	53-58	46-57	51-53	74-75
Fat (%)	19-25	18-29	19-25	7-8
Salt (%)	1.3-1.8	1-3	1.8-3.0	<1
Protein (%)	18-20	17-21	20-22	11-12
pH	5.6-6.4	6.1	5.6	6.0

40°C. Half of the whey is drained before salting, and the remainder is drained overnight in characteristic basket-weave molds without pressing. Panela, which is made from raw milk in Mexico and sold fresh and without refrigeration within 2 d of manufacture (*21*), is white, crumbly, and mild with a sweet fresh milk flavor (*7*). A variation called Queso Crema has cream added.

Queso Fresco, "fresh cheese," is coagulated with rennet, cooked at 30-32°C, drained, and salted directly or in brine (*15*). Fine milling of the curd gives the cheese a soft, crumbly texture (*14*). Queso Fresco is a mild, fresh, slightly salty cheese, which is called Adobera if it is brick-shaped (*8*). Sandra et al. conducted a study of composition and texture of Queso Fresco they prepared, finding that high-pressure treatment reduced firmness and crumbliness (*22*). Hwang and Gunasekaran also measured texture and developed a method of determining crumbliness by counting number and size of particles after crushing (*23*).

Queso Blanco, "white cheese," may be produced in two ways. Traditionally, it is acid set with lemon juice and vinegar, and without starter or rennet. Other names for this type include Queso del País and Queso de la Tierra, "cheese of the land" (in Puerto Rico) and Queso Sierra (in Mexico) (*12*). The cheese may also be coagulated with rennet, with or without the addition of mesophilic starter cultures. The curd may be cooked, and it is not milled, giving it a harder texture than Queso Fresco. The cheese is hooped and sold fresh, and has a fresh, slightly acid flavor. Although the pH of this variety is normally around 5.6, USDA purchasing specifications allow it to have a pH between 5.25 and 5.9 (*24*). Kaláb and Modler investigated Queso Blanco curd microstructure using scanning and transmission electron microscopy, finding that the compactness of the curd increased with curd temperature and duration of cooking (*25*). They also noted that renneted curd consisted of casein particles that were fused together and distinguishable, but acid coagulated curd consisted of large protein structures without distinguishable particles.

Requesón is a spreadable cheese similar to ricotta (*11*). It is made from whey collected from the production of other cheeses, with added milk or cream. Requesón is acid set with vinegar and cooked at 85°C, causing molecules of β-lactoglobulin, the primary whey protein, to denature, bind to each other, and coagulate (*10*). This variety has a mild, semisweet flavor (*8*).

Creams

Cream results from the mechanical separation of fat from skim milk. Sour cream is produced when the cream is acidified by food-grade acid or by lactic acid-producing bacteria. Literature covering the science behind Hispanic creams is scarce; more information on chemistry of cream in general may be found in McGee (26) and Early (27).

Crema Mexicana

Crema Mexicana, a fresh, slightly thick sweet cream, is similar to créme fraîche (*20*). Cream is inoculated with mesophilic bacteria, usually *Lactococcus lactis* or species of *Leuconostoc*, which sour the cream by producing lactic acid. The product contains at least 30% fat and about 3% lactose, the same as light whipping cream, and has a slightly acidic dairy flavor. The flavor arises from diacetyl, which is converted by the starter culture from citrate and which produces an intense butter flavor. Mexican families also make Crema Mexicana at home by adding buttermilk or yogurt (which also contain *L. lactis*) to heavy cream, and setting for 8 hr at room temperature and an additional 36 hr in a refrigerator (*17*). Crema Mexicana does not curdle upon heating due to its low protein content (about 2%), and it holds its shape after whipping, making it useful for cream sauces and dessert toppings. Crema Centroamericana is thicker and sweeter than Crema Mexicana, with about 50% fat and 5% lactose (*20*). It is similar to the Italian soft ripened cheese Mascarpone. A more acidic variation, Crema Centroamericana Acida, is a tangy, salted sour cream.

Crema Agria

Crema Agria, "sour cream," is thick, fresh, and pourable, has a tangy flavor, and serves as a garnish in many dishes (*20*). Cream is fermented at 20°C for 15-20 hr; fermentation is halted by rapid cooling to 5°C when the desired pH of about 4.5 is almost reached. Its acidity level of 0.8% is higher than the 0.2% of Crema Mexicana. Like sour cream in the U.S., it contains 15-20% fat and 3% protein.

Other Creams

Natas, "skin atop a liquid," consists of the rich skimmings from scalded raw milk and resembles a reduced-fat version of English clotted cream. The milk is allowed to stand for 12 hr to allow the cream to rise to the top. The milk is then heated very slowly for at least an hour, which evaporates some of the water, melts the fat, and creates a cooked flavor (17). Boiling is avoided to prevent the β-lactoglobulin from binding to casein and preventing coagulation. After cooling overnight, the surface cream, which develops the consistency of soft butter, is recovered. Natas enhances cooked sauces and some dishes (17) and can contain 50% fat.

There are other types of cream in Hispanic cuisine, but information about their chemistry is lacking. Other creams include Crema Fresca Casera ("fresh homemade cream"), a sweet, pourable whipping cream; Crema Salvadoreña, a cultured sour cream with a buttery flavor; and Crema Media, which is half-and-half.

Yogurt

In yogurt manufacture, milk, sugar, and stabilizers are combined, homogenized, and cooked at conditions ranging from 80°C for 1 hr to 92°C for 15 s. The sugar (glucose or sucrose) provide flavor, the stabilizers (such as alginates or carrageenans) prevent separation of whey from the curd, homogenization makes the product more uniform, and heat denatures β-lactoglobulin and destroys any microorganisms present in the ingredients. The denatured β-lactoglobulin adheres to κ-casein, resulting in the formation of a spongy matrix that retains whey. After cooling to 45°C, cultures of mesophilic bacteria (*Streptococcus salivarius* ssp. *thermophilus* and *Lactobacillus delbrueckii* ssp. *Bulgaricus*) are added, and the lactic acid produced causes the casein to gel in 2-3 hr. Fermentation continues to pH 4.2-4.4, and is stopped by cooling to 5°C (27). Fruit and flavors are added immediately prior to packaging; acetaldehyde generated by the *L. bulgaricus* contributes to yogurt flavor.

Cup Yogurt

Hispanic-style cup yogurt is manufactured in the same manner as in the U.S. or Europe. It is sweeter than the U.S. version due to the addition of extra sugar – up to 10% may be added (7). The fat content is 2.0-2.5%, and the product may be fortified with milk solids (7). Guava, mango, and other tropical fruits provide common Hispanic yogurt flavors. More information on the chemistry of yogurt may be found elsewhere (*12, 26*).

Drinkable Yogurt

Drinkable yogurts, also called liquid yogurt or smoothies, are more popular than regular cup yogurt in Mexico, where consumers often use them as digestive agents. Yogurt beverages are manufactured by homogenizing yogurt and then cooling and flavoring the liquid. Alternatively, regular yogurt is diluted with sucrose solution, milk, water, or fruit juice (*28*). Processing takes place under less severe conditions than cup yogurt to attain a softer curd, and more fruit is generally added (*12*). Drinkable yogurt may be made for fresh consumption or as long-life products; the latter are heat-treated after fermentation and aseptically packaged (*28*). Thin products typically contain 0.3-0.5% alginate or pectin stabilizer, while thicker products contain 0.8-2.0% starch stabilizers (*29*). The solids content does not exceed 11% and the acidity must be high enough to prevent protein coagulation (*30*). The chemistry of these products has not been reported in the literature, although a manufacturing protocol is available (*30*).

Jocoque

Jocoque, also called Labin, is a product similar to Lebanese yogurt. Its characteristics are between those of salted buttermilk and thin sour cream, and it has a tangy to sharp flavor (*17, 20*). It is made by coagulating milk with the same mesophilic bacteria used in yogurt making. The whey is drained, traditionally through a cloth bag, and cream is added after 12 hr of refrigeration (*31*). The composition is typically 9-10% fat and 22-26% total solids.

Desserts

Although desserts are popular in Hispanic cuisine, Dulce de Leche is the only dessert of Latin American origin that has undergone scientific study.

Dulce de Leche

Dulce de Leche translates to "sweet of the milk" and is often referred to as milk jam or caramel jam. In Mexico, where it is usually made from goat milk, it is also called Cajeta, "little box" (*32*). Other names are Manjar (in Chile) and

Arequipe (in Columbia and nearby countries) (*33*). There are two versions of Dulce de Leche: Casero, "homemade," a shiny, red-brown type, and Pastelero, "pastry cook," a lighter-colored bakery type with starch thickener to hold it on cakes. Both kinds are made by mixing whole milk with sucrose in a 4:1 ratio and boiling down to 70% solids (*32*). Addition of sodium bicarbonate neutralizes the mix, preventing protein coagulation (*34*). The characteristic brown color results from the Maillard reaction. The product has a sweet milk, caramel flavor. The typical composition is 50% sugar, 6% protein, and 4.5% fat (*26*), with a pH of 6.1.

Moro and Hough established standards for measuring solids and density using refractometry (*32*). Hough et al. measured flow properties at varying solids contents and temperatures (*35*), and Rovedo et al. found that the apparent viscosity of Casero increased with alkalinity and storage time, and that Pastelero had more consistency because of the added starch (*34*). Both had thixotropic (lower viscosity when mechanically disturbed) and pseudoplastic (lower viscosity with increasing shear rate) characteristics (*34*).

Other Desserts

Ice cream is a popular Hispanic dessert, and is made the same way as in much of the world. Milk, cream, milk powder, sugar, polysaccharide stabilizer, and glyceride emulsifier are pasteurized and homogenized, flavor is added, and the mix is whipped to introduce air and then chilled to -30°C or under (*27*). Flavors not normally seen in the U.S. include banana, coconut, eggnog, lime, pineapple, and rose petal. Mexican fried ice cream is a relatively recent invention along the lines of baked Alaska, in which a scoop of ice cream is rolled in sweetened corn flakes and browned in an oven (*36*). The chemistry of ice cream has been detailed by Marshall et al. (*37*).

Tres Leches, "three milks," is a combination of cream, sweetened condensed milk, and evaporated milk and is usually poured on cake. After baking, the cake is perforated and soaked in the mixture until moist but not mushy (*35*). It is also used as an ice cream flavor (*6*).

Summary

The consumption of Hispanic dairy products in the U.S. will continue to grow with their population. An understanding of the characteristics of these products is essential if their properties are to be optimized and their quality is to be improved. It is hoped that this chapter provides part of the foundation for the research needed to reach this goal.

References

1. Ramirez, R.R.; de la Cruz, G.P. The Hispanic Population in the United States: March 2002, Current Population Reports, U.S. Census Bureau, Washington, DC, 2003.
2. Ennen, S. *Food Proc.* **2003,** *64*(9), 52-56.
3. Decker, K.J. *Food Prod. Design* **2004,** *14*(6), 35-60.
4. Romero-Gwynn, E.; Gwynn, D. Dietary patterns and acculturation among Latinos of Mexican descent, JSRI research report #23. Michigan St. Univ., E. Lansing, MI, 1997.
5. National Agricultural Statistics Service. Dairy Products 2004 Summary. USDA, Washington, DC, 2005.
6. Hutchinson, M.A. A flavor-inspired renaissance for dairy. *Prepared Foods* Web Site, www.preparedfoods.com/CDA/ArticleInformation/ features/BNP__Features__Item/0,1231,113498,00.html. Posted Nov. 26, 2003.
7. Food and Agriculture Organization of the United Nations. The Technology of Traditional Milk Products in Developing Countries. FAO, Rome, Italy, 1990.
8. Van Hekken, D.L.; Farkye, N.Y. *Food Technol.* **2003,** *57*(1), 32-38.
9. Berry, D. 2003 cultured product trends: The growth continues. *Dairy Foods* Web Site, www.dairyfoods.com/CDA/ArticleInformation/ coverstory/BNPCoverStoryItem/0,6809,109387,00.html. Posted Oct. 7, 2003.
10. Lucey, J.A.; Johnson, M.E.; Horne, D.S. *J. Dairy Sci.* **2003,** *86,* 2725-2743.
11. Jenkins, S. Cheese Primer. Workman Publ., New York, 1996, p. 392.
12. Kosikowski, F.V.; Mistry, V.V. Cheese and Fermented Milk Foods. Principles and Practices, Vol. 1. 3rd edn. F.V. Kosikowski LLC, Westport, CT, 1997.
13. Phelan, J.A.; Renaud, J.; Fox, P.F. Cheese: Chemistry, Physics and Microbiology. Major Cheese Groups, Vol. 2. Chapman and Hall, London, 1993, pp 460-461.
14. Nauth, R. Fundamentals of Practical Cheese Manufacture. Institute of Food Technologists, Chicago, IL, 2004.
15. Path, J. *UW Dairy Pipeline* **1991,** *3*(4), 1-4.
16. De Alba, L.A.; Staff, C.; Richter, R.L.; Dill, C.W. *Cult. Dairy Prod. J.* **1991,** *26*(2), 11-12.
17. Kennedy, D. From My Mexican Kitchen: Techniques and Ingredients. Clarkson Potter Publ., New York, pp 22-29.
18. Tunick, M.H.; Van Hekken, D.L.; Molina-Corral, F.J.; Tomasula, P.M.; Call, J.E.; Luchansky, J.B.; Gardea, A.A. *Int. J. Dairy Technol.* Submitted 2006.

19. Tunick, M.H.; Van Hekken, D.L.; Call, J.E.; Molina-Corral, F.J.; Gardea, A.A. *Int. J. Dairy Technol.* Submitted 2006.
20. Alden, L. The Cook's Thesaurus Web Site, www.foodsubs.com/Cultmilk.html. Updated 2005.
21. Saltijeral, J.A.; Alvarez, V.B.; Garcia, B. *J. Food Safety* **1999**, *19*, 241-247.
22. Sandra, S.; Stanford, M.A.; Meunier Goddik, L. *J. Food Sci.* **2004**, *69*, 153-158.
23. Hwang, C.H.; Gunsasekaran, S. *Milchwissenschaft* **2001**, *56*, 446-450.
24. USDA, Agricultural Marketing Service. Commercial Item Description. Cheese, Queso Blanco. www.ams.usda.gov/fqa/aa20347.htm. Posted Dec. 20, 2004.
25. Kaláb, M.; Modler, H.W. *Food Microstruct.* **1985**, *4*, 89-98.
26. McGee, H. *On Food and Cooking. 2nd edn.* Scribner, New York, 2004.
27. Early, R. *The Technology of Dairy Products.* VCH Publ., New York, 1992.
28. Kurmann, J.A.; Rašić, J.L.; Kroger, M. *Encyclopedia of Fermented Fresh Milk Products.* Van Nostrand Reinhold, New York, NY, p. 315.
29. Anonymous. Catching the wave of drinkable yogurts. *Dairy Foods* Web Site, www.dairyfoods.com/CDA/ArticleInformation/coverstory/BNP Cover Story Item/0,6809,109387,00.html. Posted Sept. 1, 2003.
30. Washington Red Raspberry Commission. Red Raspberries in Dairy Product Formulations. www.red-raspberry.org/PDF/DairyMono.pdf. Posted 2005.
31. Agroterra.com. Jocoque Natural, www.agroterra.com/mercado/det_sector.ASP?IdProducto=5096. Updated June 28, 2005.
32. Moro, O.; Hough, G. *J. Dairy Sci.* **1985**, *68*, 521-525.
33. Smith, K.; Burrington, K.J. *UW Dairy Pipeline* **1999**, *11*(2), 10.
34. Rovedo, C.O.; Viollaz, P.E.; Suarez, C. *J. Dairy Sci.* **1991**, *74*, 1497-1502.
35. Hough, G.; Moro, O.; Segura, J.; Calvo, N. *J. Dairy Sci.* **1988**, *71*, 1783-1788.
36. Olver, L. The Food Timeline Web Site, www.foodtimeline.org/food icecream.html. Updated Dec. 31, 2004.
37. Marshall, R.T.; Goff, H.D.; Hartel, R.W. *Ice Cream. 6th edn.* Plenum Press, New York, 2003.

Chapter 4

Chemical Characterization of *Lippia graveolensi* Kunth and Comparison to *Origanum vulgare* and *Origanum laevigatum* 'Herrenhausen'

Salvador Lecona-Uribe[1], Guadalupe Loarca-Piña[1],
Cynthia Arcila-Lozano[1], and Keith R. Cadwallader[2]

[1]PROPAC, Research and Graduate Studies in Food Science, School of
Chemistry, University of Querétaro, Qro., Mexico 76010
[2]Department of Food Science and Human Nutrition, University of Illinois at
Urbana-Champaign, 1302 West Pennsylvania Avenue, Urbana, IL 61801

Oregano includes several genera and species, the most
common being European *Origanum vulgare*, and Mexican
Lippia graveolens. The chemical composition of essential oils
and aqueous extracts from Mexican oregano (*Lippia
graveolens* Kunth), air-dried with or without exposure to
sunlight, was determined. The main essential oil components
were thymol (66%) and carvacrol (1.9%) and their precursors,
p-cymene (10%) and γ-terpinene (2.8%). Total polyphenol
content of the aqueous extract of *L. graveolens* was 656 μg eq.
EGCG/g dry leaves. Volatile composition was affected by sun-
drying which lead to the formation of oxidized derivatives.
Mexican oregano is a rich source of flavor and phenolic
compounds.

45

Several plant species and their extracts have been a major source of research for the investigation of their biological properties, such as antioxidant, antimicrobial, antimutagenic, and anticarcinogenic activites (*1-4*). These biological properties may be due to the presence of phytochemicals including phenolic compounds, phenolic acids, and terpenoids (*5-7*).

The oregano herb includes various plant species. The most common are the genera *Origanum* native to Europe and *Lippia* native to Mexico (*8*). The phytochemical analysis of dried leaves and essential oils of *Origanum* sp. showed that the main volatile components are limonene, β-caryophyllene, p-cymene, camphor, linalool, α-pinene, carvacrol, and thymol (*9, 10*). The volatile composition of oregano depends on the species and cultivation practices such as climate and altitude, time of collection, and stage of growth. Some of the properties of this plant's extract are currently under investigation due to growing interest in suitable natural substitutes for synthetic additives commonly found in foods. For example, essential oils and some extracts from *Origanum acutidens* showed better antioxidant activity than butylated hydroxytoluene (BHT) and greater antimicrobial activity against *Escherichia coli* O157:H7 and *Salmonella enterica* in apple juice (*11*). Oussalah et al. (*1*) reported the use of edible films containing oregano essential oil and pimento as a good preservation method for meat.

The objective of the present study was to determine the chemical composition of essential oils and aqueous extracts from Mexican oregano (*Lippia graveolens* Kunth), air-dried with or without exposure to sunlight.

Materials and Methods

Chemicals

Epigallocatechin gallate, (+)-catechin, gallic acid, theobromin, gallocatechin, syringic acid, myrcetin, naringenin, pinocembrin, monoterpenes, carvacrol, γ-tepinene, p-cymene, thymol, naringenin, and pinocembrin as well as volatile terpenes and sequiterpenes indicated in Figures 1 and 2 and Table 1 were obtained from Sigma-Aldrich (St. Louis, MO).

Oregano Leaves and Essential Oil

Dried leaves of *Lippia graveolens* Kunth (*L. graveolens*) were kindly provided by Facultad de Biología and the essential oil was from Facultad de

Ciencias Químicas de la Universidad Autónoma de Querétaro. Essential oil was also extracted by hydrodistillation of the plant material (leaf only, 35 g) with 500 mL distilled deionized water (ddH$_2$O). The distillation period was 3 hr, and the essential oil obtained was stored under refrigeration (4°C) in an amber vial until GC-MS analysis.

Dried leaves of *Origanum vulgare* (*O. vulgare*), *Origanum laevigatum* 'Herrenhausen' (*O. laevigatum*) and *Lippia graveolens* were grown and harvested in a greenhouse at the University of Illinois at Urbana-Champaign.

Preparation of Herbal Tea

Fresh leaves of *O. vulgare* and *O. laevigatum* were air-dried without exposure to sunlight, while leaves of *L. graveolens* were dried with or without exposure to sunlight. The aqueous extracts of *O. vulgare, O. laevigatum,* and *L. graveolens* were prepared as described previously (*12*). The dried leaves (DL) were kept in plastic bags and refrigerated at 4°C. The extract was prepared from 2.7 g DL that was soaked in 250 mL of boiling water (98°C) for 10 min. The mixture was cooled to room temperature before filtration using a 0.45-µm filter and then lyophilized. The freeze-dried solid extract was kept at -20°C in plastic tubes, sealed with parafilm and protected from light. Solid extracts were redissolved in ddH$_2$O and filtered with using a 0.22-µL syringe filter prior to analysis. Each preparation was standardized for total polyphenol content.

Total Polyphenol Content

The total polyphenol content of the freeze-dried teas were measured as described by the modified Folin-Ciocalteu method (*13*). Briefly, 1 mL of 1 N Folin-Ciocalteu reagent was added to a 1 mL sample, and this mixture was allowed to stand for 2-5 min before the addition of 2 mL of 20% Na$_2$CO$_3$. The solution was then allowed to stand for 10 additional minutes before reading the absorbance at 730 nm in a Beckman DU 640 spectrophotometer (Beckman Coulter, Fullerton, CA). The total polyphenol content was expressed as milligram equivalents to the standard used per milliliter of aqueous extract or gram of DL. Equations obtained for standard curves were $y = 0.0125x - 0.0758$, $r^2 = 0.992$; $y = 0.0267x - 0.0966$, $r^2 = 0.993$; and $y = 0.0269x - 0.0500$, $r^2 = 0.983$; for EGCG; gallic acid; and (+)-catechin, respectively.

Simultaneous Distillation Extraction (SDE)

Five grams of *L. graveolens* dried leaves (air-dried with or without exposure to sunlight) in 40 mL of ddH$_2$O were placed in a 100-mL round-bottom flask and extracted in a micro SDE apparatus with 10 mL of diethyl ether. The distillation period was 2 hr and the ether extract obtained was stored at 4°C, light protected, until GC-MS analysis.

Partial Characterization of Teas

The characterization of phenolic compounds was performed using a 1050 series gradient liquid chromatograph (Agilent Technologies, Inc., Palo Alto, CA), equipped with a 1050 auto sampler, a 1050 gradient pump, a 1050 photodiode array detector (PDA), and helium sparge. Separations were performed using a C$_{18}$ RP guard column and a C$_{18}$ RP Prodigy ODS column (250 mm x 4.6 mm i.d. x 5 μm; Phenomenex, Torrance, CA). The column temperature was ambient, and the elution (1 mL/min) was performed using a solvent system comprising of solvents A (water/methanol/formic acid, 79.7/20/0.3) and B (methanol/formic acid, 99.7/0.3) mixed using a gradient starting with 100% A, linearly decreasing to 48% A in 52 min, to 20% A in 5 min, and held at 20% A for 3 min, then a linear increase to 100% A in 5 min. The initial condition was held for 5 min before the next run. The injection volume was 5-50 μL. Sample concentration was 0.3 mg/mL for pure standards and 0.84 μg equivalent of (+)-catechin. The PDA detector was set in the range of 195-450 nm, with outputs at 260, 280, 330, and 360 nm. The partial identification of phenolic compounds was based on standard retention times, spike standards, and spectra comparisons. A spectral library was built up from 25 of the standards mentioned in the chemicals section. Calibration curves were generated for GA, (+)-catechin and EGCG by plotting concentrations versus peak areas. Regression equations were used to calculate percentage recovery for GA ($y = 2020.96x + 5.51$, $r^2 = 0.999$) and EGCG ($y = 1423.47x - 7.12$, $r^2 = 0.999$).

Gas Chromatography-Mass Spectrometry (GC-MS)

The GC-MS system consisted of an Agilent 6890 GC/5973 mass selective detector (MSD; Agilent Technologies, Inc.). Samples were injected using cool on-column mode into an HP5-MS (30 m x 0.25 mm i.d. x 0.5μm film; Agilent Technologies, Inc.) or SPB-50 (30 m x 250 μm i.d. x 0.25 μm film; Supelco, Bellefonte, PA) column. Helium was the carrier gas at a constant flow of 1 mL/min. The GC oven temperature was programmed from 50 to 225°C at a rate of 4°C/min with initial and final hold times of 5 and 83.33 min, respectively. MSD conditions were as follows: capillary direct interface temperature, 280°C; ionization energy, 70 eV; mass range, 35-300 amu; electron multiplier voltage (Autotune + 200 V); scan rate, 5.27 scans/s.

Compounds were identified by comparison of their retention times and mass spectra against authentic reference compounds. Tentative identifications were based on comparison the mass spectra to a standard library (Wiley 138K Mass Spectral Database, Wiley and Sons, 1990).

Results and Discussion

Characterization of *L. graveolens*

The main components of *L. graveolens*. were thymol (66.12%), *p*-cymene (11.36%), α-terpinene (2.82%), carvacrol (2.38%), β-myrcene (2.38%), and caryophyllene (1.9%) (Fig. 1). The results agree with the reported composition of oregano (*14, 15*) with minor differences attributed to variations between oregano species and growth conditions.

In order to learn about the main chemical components of the dried leaves of *L. graveolens*, an ether extract was obtained from leaves that were air-dried with or without exposure to sunlight (Fig. 2). As expected, additional components were found that are not normally present in the essential oil (Table I). The main compound was thymol (10.18 ± 3.17% and 4.67 ± 0.87% in air-dried leaves prepared with or without exposure to sunlight, respectively). The volatile composition was affected by sun-drying which lead to the formation of oxidized derivatives. For example, the amount of thymol was reduced by two-fold under sun-drying conditions.

Characterization of *O. vulgare* and *O. laevigatum*

A preliminary characterization of phenolic compounds in the aqueous extract of *O. vulgare* and *O. laevigatum* is shown in Figure 3. Alves Rodriguez et al. (*16*) reported thymol and cis-sabinene hydrate as a major components in *O. vulgare* dried leaves, using sub- and critical CO_2 extraction methods. These compounds were observed by GC-MS in the present study. Additional phenolic components were identified by HPLC (Fig. 3).

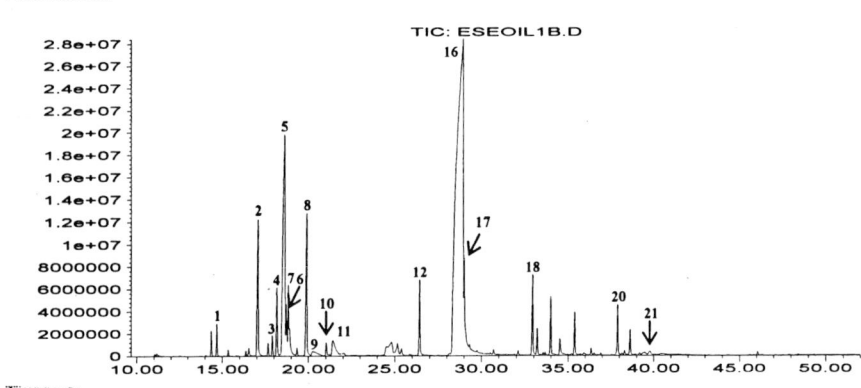

Figure 1. GC-MS chromatogram of the volatile components of Lippia graveolens Kunth essential oil. Peak identifications: 1 = α-pinene, 2 =β-myrcene, 3 = δ-3-carene, 4 = α-terpinene, 5 = p-cymene, 6 = limonene, 7 = eucalyptol, 8 = γ-terpinene, 9 = sabinene hydrate, 10 = α-terpinolene, 11 = linalool, 12 = methyl thymyl ether, 16 = thymol, 17 = carvacrol, 18 =β-caryophyllene, 20 = caryophyllene oxide, 21 = α-/β-eudesmol.

Total Polyphenolic Content

As illustrated in Tables II and III, total polyphenolic content in aqueous extracts was higher in *Origanum* than in *Lippia*. Values are expressed as µg eq. EGCG/ml, µg eq. gallic acid/mL, or µg eq. catechin/mL, and as µg/g dry leaves. *O. laevigatum* had higher values for EGCG and gallic acid (656 ± 20 µg eq. EGCG/g DL and 360 ± 10 µg eq. gallic acid/g DL, respectively) and *L. graveolens* dried under sunlight conditions had the lowest levels (217 ± 5 µg eq. EGCG/g DL, 1.29 ± 0.18 µg eq. gallic acid/g DL and 94 ± 2 µg eq. (+)-cathechin/g DL). Statistical difference was observed between *O. laevigatum* and *O. vulgare* for EGCG and (+)-catechin (expressed as µg eq. of phenolic compound/g DL), and no statistical difference was observed between *L. graveolens* leaves dried under non-sunlight and sunlight conditions. Sökmen et al. (*11*) reported 1.9 ± 0.4 µg eq gallic acid/mg in hexane extract from *O. acudiens*. Aqueous extract from *O. vulgare* and methanolic extract from *O. acudidens* have been shown to have biological activities, such as anti-hypoglycaemic and antiviral activities, respectively (*7, 11*).

Table I. Main Volatile Constituents in Ether Extract of *L. Graveolens* Dried under Sunlight and Non-Sunlight Conditions (%)

Component	Non-sunlight	Sunlight
α-Pinene	1.0 ± 0.03	0.97 ± 0.3
β-Myrcene	4.49 ± 0.24	2.46 ± 1.16
δ-3-Carene	5.72 ± 0.24	4.66 ± 1.21
α-Terpinene	2.10 ± 0.08	1.52 ± 0.44
p-Cymene	7.54 ± 0.51	8.24 ± 2.49
Limonene	3.03 ± 0.15	2.41 ± 0.67
Eucalyptol	10.53 ± 0.7	12.28 ± 2.76
γ-Terpinene	6.3 ± 0.23	5.11 ± 1.15
Sabinene hydrate	1.86 ± 0.03	1.65 ± 0.09
α-Terpinen-4-ol	3.25 ± 0.09	2.13 ± 0.39
Linalool	7.26 ± 0.3	7.8 ± 1.17
Borneol	1.54 ± 0.06	1.13 ± 0.36
Terpinen-4-ol	7.76 ± 0.4	8.46 ± 1.76
α-Terpineol	3.16 ± 0.04	3.84 ± 0.93
Thymol	4.67 ± 0.87	10.18 ± 3.17
Carvacrol	0.22 ± 0.02	0.49 ± 0.23
β-Caryophyllene	3.25 ± 0.14	3.1 ± 0.79
α-Caryophyllene	2.01 ± 0.13	1.95 ± 0.79
Caryophyllene oxide	1.43 ± 0.08	1.74 ± 0.53
α-/β-Eudesmol*	1.45 ± 0.22	2.21 ± 0.77

Results represent the average of two independent experiments with three replicates ± SD.
* = compound tentatively indentified.

Table II. Total Polyphenolic Content of the Aqueous Extracts of *Oreganum laevigatum* and *Oreganum vulgare*

Phenolic compounds (eq)	Origanum laevigatum		Origanum vulgare	
	μg/mL	μg/g DL	μg/mL	μg/g DL
EGCG	9.23 ± 0.67[a]	656 ± 20[b]	7.64 ± 0.1[c]	600 ± 6[d]
Gallic acid	5.10 ± 0.32[a]	360 ± 10[b]	4.36 ± 0.04[c]	350 ± 3[b]
(+) Catechin	3.33 ± 0.31[a]	233 ± 10[b]	2.59 ± 0.04[c]	300 ± 3[d]

Results represent the average of two independent experiments with three replicates ± SD.
Means with different superscript letters in the same line are significantly different (Tukey $\alpha = 0.05$). EGCG = Epigallo catechin gallate.

52

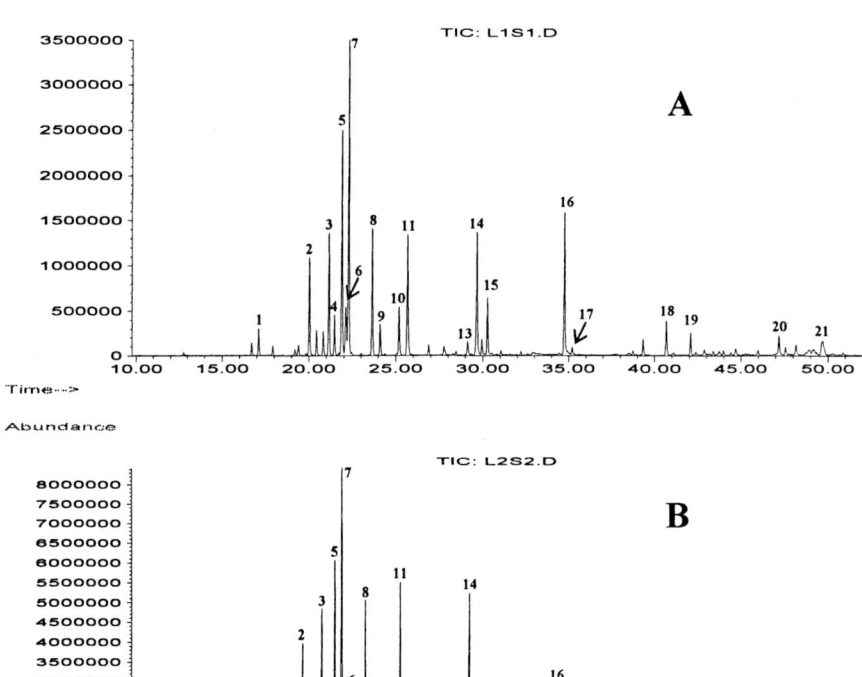

Figure 2. GC-MS comparision of the volatile components of ether extracts from Lippia graveolens Kunth leaves dried under non-sunlight (A) and sunlight (B) conditions. Peak identifications: 1 = α-pinene, 2 = β-myrcene, 3 = δ-3-carene, 4 = α-terpinene, 5 = p-cymene, 6 = limonene, 7 = eucalyptol, 8 = γ-terpinene, 9 = sabinene hydrate, 10 = α-terpinolene, 11 = linalool, 13 = borneol, 14 = terpinen-4-ol, 15 = α-terpineol, 16 = thymol, 17 = carvacrol, 18 = β-caryophyllene, 19 = α-caryophyllene, 20 = caryophyllene oxide, 21 = α-/β-eudesmol.

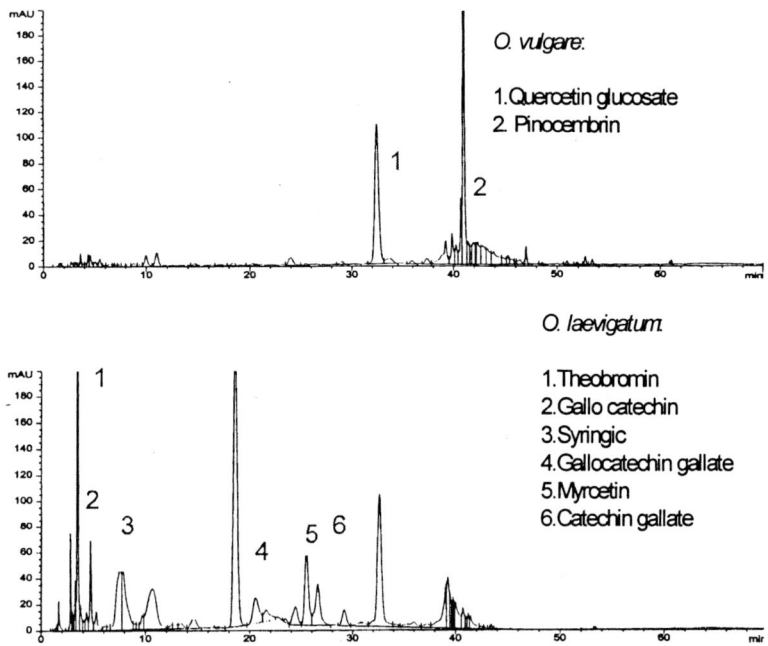

Figure 3. HPLC chromatogram of the aqueous extracts from Origanum vulgare and Origanum laevigatum 'Herrenhausen.'

Table III. Total Polyphenol Content of the Aqueous Extracts of *Lippia graveolens* Kunth

Phenolic compounds (eq)	Non-sunlight		Sunlight	
	$\mu g/mL$	$\mu g/g\ DL$	$\mu g/mL$	$\mu g/g\ DL$
EGCG	2.82 ± 0.16^a	223 ± 5^b	2.86 ± 0.07^a	217 ± 5^b
Gallic acid	1.39 ± 0.07^a	110 ± 6^b	1.29 ± 0.18^a	108 ± 2^b
(+) Catechin	1.21 ± 0.07^a	96 ± 6^b	1.23 ± 0.30^a	94 ± 2^b

Results represent the average of two independent experiments with three replicates ± SD. Means with different superscript letters in the same line are significantly different (Tukey $\alpha = 0.05$). EGCG = Epigallo catechin gallate.

Conclusions

The volatile composition of Mexican oregano (*L. graveolens* Kunth) can be affected by sun-drying leading to the formation of oxidized derivatives. Thymol is reduced by half under sun-drying conditions (4.7% and 10.2% DL without sun exposure and with sun exposure, respectively), as well as with canvacrol content (0.2% and 0.5% DL without sun exposure and with sun exposure, respectively). Mexican oregano is a reach source of flavor compounds that can be used as additives in the food industry. Based on its composition, this herb could be commercially exploited and compete favorably with European or African oregano (*Oreganum* sp.).

Acknowledgments

This study was made possible by the generous support from the United States Agency for International Development through a grant from the Association Liaison Office for University Cooperation in Development. The support provided to Visiting Faculty of Guadalupe Loarca-Pina by International Programs, University of Illinois is also acknowledged.

References

1. Oussalah, M.; Caillet, S. ; Salmiera, S.; Saucier, L. ; Lacroix, M. *J. Agric. Food. Chem.* **2004**, *52,* 5598-5605.
2. Matsuura, J.; Chihi, H.; Asakawa, C.; Amano, M.; Yoshihara, T.; Mizutani, J. *Biosci. Biotechnol. Biochem.* **2003**, *67(11),* 2311-2316.
3. He, L.; Mo, H.; Hadisusilo, S.; Qureshi, A.A.; Elson, C.E. *J. Nutr.* **1997**, *127,* 668-674.
4. Kanazawa, K.; Kawasaki, H.; Samejima, K.; Ashida, H.; Danno, G. *J. Agric. Food Chem.* **1995**, *45*, 404-409.
5. D'antuono, L. F.; Galletti, G.C.; Bocchini, P. *Ann. Bot.* **2000**, *86*, 471-478.
6. Hernández, T.; Canales, M.; Avila, J.G.; Duran, A.; Caballero, J.; Romo de Vivar, A.; Lira, R. *J. Ethnopharmacol.* **2003**, *88*, 181-188.
7. Friedman, M.; Henika, P.R.; Levin, C.E.; Mandrell, R.E. *J. Agric. Food Chem.* **2004**, *52,* 6042-6040.
8. *Practical Guide to Natural Medicines.* Pierce, A., Ed.; William Morrow and Company: New York, 1999; p 728.
9. Pascual, M.E.; Slowing, K.; Carretero, E.; Sánchez Mata, D.; Villar, A. *J. Ethnopharmacol.* **2001**, *76*, 201-214.

10. Nostro, A.; Blanco, A.R.; Cannatelli, M.A.; Enea, V.; Flamini, G.; Morelli, I.; Roccaro, A.S.; Alonzo, V. *FEMS Microbiol. Lett.* **2004**, *230,* 191-195.
11. Sökmen, M., Serkedjieva, J.; Daferera, D.; Gulluce, M.; Polissiou, M.; Tepe, B.; Akpulat, A.; Sahin, F.; Sökmen, A. *J. Agric. Food. Chem.* **2004**, *52,* 3309-3312.
12. González de Mejía, E.; Ramírez-Mares, M.V.; Nair, M.G. *J. Agric. Food Chem.* **2002**, *50,* 7714-7719.
13. Nurmi, K.; Ossipov, V.; Haukioja, E.; Pihlaja, K. *J. Chem. Ecol.* **1996**, *22,* 2023-2040.
14. Exarchou, V.; Godejohann, M.; van Beek, T.A.; Gerothanassis, I.P.; Vervoort, J. *Anal. Chem.* **2003**, *75,* 6288-6294.
15. Velluti, A.; Sanchis, V.; Ramos, A.J.; Egido, J.; Marín, S. *J. Food Toxicol.* **2003**, *89,* 145-154.
16. Alves Rodriguez, M.R.; Canielas, K.L.; Basto, C.M.; Dos Santos, J.G.; Dariva, C.; Vladimir, J.O. *J. Agric. Food Chem.* **2004**, *52,* 3042-3047.

Chapter 5

Aroma Components of Chipotle Peppers

Keith R. Cadwallader, Theresa A. Gnadt, and Lizzet Jasso

Department of Food Science and Human Nutrition, University of Illinois at
Urbana-Champaign, 1302 West Pennsylvania Avenue, Urbana, IL 61801

Chipotle peppers, known for their spicy and smoky flavor, are
a popular ingredient in Latin American cuisine. Three
authentic commercial samples (Mexican origin) of dried
chipotle peppers were obtained from specialty foods retailers.
Volatiles were isolated by direct solvent and extraction-
solvent-assisted flavor extraction. Aroma impact compounds
were identified by gas chromatography-olfactometry (GCO)
and GC-MS. Characterizing aroma compounds included
compounds inherent to fresh ripe jalapeno peppers (e.g. 2-
isobutyl-3-methoxyphenol, linalool, and several lipoxygenase-
derived compounds) and those imparted by smoking (e.g.,
phenol, guaiacol, and syringol derivatives) and drying (via
Maillard/Strecker reactions, e.g. methylpropanal, 2,3-
butanedione, 2- and 3-methylbutanal, 2- and 3-methylbutanoic
acid, 4-hydroxy-2,5-dimethyl-3($2H$)-furanone, and 3-hydroxy-
4,5-dimethyl-2($5H$)-furanone).

Chipotle peppers, developed centuries ago in the region north of present-day
Mexico City, are Jalapeño peppers (*Capsicum annuum var. annuum*) that have
been smoked and dried. Chipotle peppers are an important ingredient in Mexican
cooking and have become increasingly popular in Latin American cuisine since
their introduction in the 1950's. Despite their popularity, no studies have been
published on the characteristic aroma constituents of chipotle peppers and only

© 2007 American Chemical Society   **57** limited research has been conducted on Jalapeño pepper volatile constituents. Huffman et al. (*1*) reported that the characterizing flavor of Jalapeño peppers was due to 2-isobutyl-3-methoxypyrazine, which is also a key bell pepper aroma component. Analysis of three *Capsicum annuum* cultivars, Jalapeño, Anaheim, and Fresno, showed that among the three cultivars, Jalapeño peppers had the widest range of volatiles (*2*).

The past two decades have experienced tremendous growth in the development and use of gas chromatography-olfactometry (GCO) for the identification and characterization of the aroma-active constituents of foods. There are numerous techniques available to collect, analyze and record GCO results included aroma extract dilution analysis (AEDA), CharmAnalysis, cross modal modeling, frequency of detection, and post-peak intensity scaling (*3, 4*). Among these, post-peak intensity scaling provides a simple and rapid way to evaluate the odor quality and intensity of individual aroma-active compounds in a complex aroma extract. The quality of the GCO results, however, depends greatly on whether or not the aroma extract is truly representative of the original food product. The combined use of direct solvent extraction (DSE) under mild conditions with a subsequent mild cleanup step, such as solvent-assisted flavor evaporation (SAFE) under high vacuum, provides for the careful and accurate isolation of food volatiles with little or no volatile profile alteration due to formation of artifacts or loss of labile constituents (*5*).

The aim of the present study was to apply DSE-SAFE and GCO with post-peak intensity scaling for the identification and characterization of the aroma-active components of commercial dried chipotle peppers.

Experimental Procedures

Materials

Three authentic commercial samples (Mexican origin) of dried chipotle peppers were obtained from specialty foods retailers. Samples were stored away from light at room temperature 23-25° C in their original packaging until analyzed.

Analytical grade diethyl ether and reference compounds listed in Table I were obtained from Sigma-Aldrich, Inc. (St. Louis, MO), except for compound no. 7 from Lancaster Synthesis, Inc. (Windham, NH), no. 26 from Firmenich Inc. (Princeton, NJ), and nos. 8 and 31 were synthesized using published procedures (*6, 7*), respectively. Odor-free water was prepared by boiling deionized-distilled water in an open flask until its volume was decreased by one-third.

Isolation of Volatiles

Direct Solvent Extraction (DSE)

De-stemmed chipotle peppers were ground in an Osterizer 4096 (Niles, IL) commercial blender and sieved through a No. 18 U.S.A. Standard Testing Sieve (W.S. Tyler, Inc., Mentor, OH). Twenty-five grams of chipotle powder and 100 mL of odor-free water were placed in a 250-mL PTFE screw-capped bottle (Nalge Company, Rochester, NY). In order to fully hydrate the powder the bottle was shaken for 30 min at 200 rpm on an orbital shaker (VWR Scientific Products, West Chester, PA). Fifty milliters of diethyl ether were then added to the hydrated sample and the bottle was shaken (20 min at 200 rpm) and then centrifuged (3000xg for 4 min on an IEC HC-SII centrifuge; Damon/IEC Division, Ramsey, MN) to break the emulsion. The solvent layer was recovered and the extraction procedure was repeated two more times using 25 mL of ether in each extraction. The pooled solvent extract was frozen overnight at -70°C to remove bulk water as ice crystals.

Solvent-Assisted Flavor Evaporation (SAFE)

SAFE was conducted on the solvent extract in order to isolate the volatile constituents away from any nonvolatile components present in the extract. This was an important step since on-column injection was used during GC analysis. Solvent extracts were subjected to SAFE as earlier described (5, 8). The "aroma extract" from SAFE was concentrated to 20 mL on a Vigruex column at 42°C, dried over 1 g of anhydrous sodium sulfate and then further concentrated under a gentle stream of nitrogen to 0.5 mL. One aroma extract was prepared for each sample. Extracts were stored in glass vials at -70°C until analysis.

Analysis of Aroma Extracts

Gas Chromatography-Olfactometry (GCO) and GC-Mass Spectrometry (MS)

Aroma-active constituents were determined by GCO and GC-MS with cool on-column injection (9) and GCO-post-peak intensity scaling was done using a seven point scale, where 0 = not detected and 7 = very strong.

Compound Identification

Positive identifications of aroma compounds were made by matching GC retention indices (RI) (10) on polar and non-polar columns, mass spectra, and aroma properties of unknowns with those of authentic reference standards analyzed under identical conditions. Tentative identifications were based on standard MS library comparison (Wiley 138K Mass Spectral Database, Wiley and Sons, 1990) or on matching RI values with authentic compounds or published literature.

Results and Discussion

Predominant Odorants in Chipotle Peppers

A total of 41 odorants were detected by GCO of DSE-SAFE aroma extracts prepared from the three dried chipotle pepper samples (Table I). All but five compounds (nos. **5, 10, 13, 15** and **31**) were detected in all three samples. Thirty-four compounds were positively identified, seven (nos. **7, 8, 10, 15, 23, 26, 31**) were tentatively identified and one compound (no. **24**) was not identified. The odorants could be grouped based on the average odor intensities observed for the three samples. Sixteen compounds having the highest odor intensities ($\geq$ 4.0) were considered as key aroma components of chipotle peppers. These included 2-methylbutanal (no. **2**), 3-methylbutanal (no. **3**), 2-ethyl-3,5-dimethylpyrazine (no. **12**), 2-isobutyl-3-methoxypyrazine (no. **14**), 2- and 3-methylbutanoic acid (coeluted as peak no. **21/22**), β-damascenone (no. **26**), guaiacol (no. **27**), o-cresol (no. **30**), Furaneol (no. **32**), octanoic acid (no. **34**), p-cresol (no. **35**), sotolon (no. **37**), syringol (no. **38**), coumarin (no. **39**), phenyacetic acid (no. **41**), and vanillin (no. **42**). Seven additional odorants may also be important in overall chipotle pepper aroma (odor intensities $\geq$ 3), including 1-octen-3-one (no. **7**), dimethyltrisulfide (no. **9**), acetic acid (no. **11**), linalool (no. **16**), butanoic acid (no. **19**), 2-phenylethanol (no. **28**), 4-methylguaiacol (no. **29**), and skatole (no. **40**).

Aroma Chemistry of Chipotle Peppers

The aroma constituents of chipotle peppers can be subdivided into three general categories based on the possible origins. Some compounds appear to be inherent to fresh Jalapeño peppers (e.g. nos. **10, 14, 16, 18, 25**), while others could be formed as a result of heating/drying processes (e.g. nos. **1-4, 17, 21/22, 26, 32, 34, 37, 40, 41**) or from wood smoking (e.g. nos. **27, 29, 30, 33, 35, 38**).

It is likely that some compounds could possibly originate from more than one source.

Compounds Inherent to Fresh Jalapeño Peppers

Buttery et al. (*11*) indicated the importance 2-isobutyl-3-methoxypyrazine (no. **14**), (*E,Z*)-2,6-nonadienal (no. **18**), and (*E,E*)-2,4-decadienal (no. **25**) in bell pepper aroma. Linalool (no. **15**) was also identified by these researchers as a bell pepper volatile constituent. Huffman et al. (*1*) later reported the occurrence of 2-isobutyl-3-methoxypyrazine in Jalapeño pepper, which was suggested by the authors to be a key component of its flavor. Luning et al. (*12*) detected 30 odorants by GCO in bell pepper, among which only nine were identified. Notable among these were 2-isobutyl-3-methoxypyrazine, linalool and 2,3-butanedione (no. **4**) which were all identified in chipotle peppers in the present study. In addition to the above, the earthy smelling odorant 2-isopropyl-3-methoxypyrazine (no. **11**) has also been previous reported in bell pepper (*13, 14*).

Short-chain (C_6) aldehydes, alcohols and ketones, which are often associated with fresh-cut peppers, were notably absent in the present study. On the other hand, medium-chain (C_8, C_9, and C_{10}) aldehydes and ketones were identified as aroma-active components of chipotle peppers in our study. The involvement of lipoxygenase (LOX) and fatty acid hydroperoxide lyase (HPL) enzymes in the formation of volatile aldehydes, alcohols and ketones has been reported in bell pepper (*12, 15*). Wu and Liou (*16*) demonstrated that certain compounds such as volatile aldehydes and alcohols increased as a result of tissue disruption, while others such as terpenes (e.g. linalool) and 2-isobutyl-3-methoxypyrazine showed no appreciable change in concentration. Occurrence of the C_9 aldehyes (nos. **16** and **18**) and C_8 ketones (nos. **7** and **8**) could be explained by the action of LOX and HPL enzymes.

Among the various compounds inherent to fresh Jalapeno peppers, 2-isobutyl-3-methoxypyrazine (no. **14**) and linalool (no. **15**) seem to be important odorants contributing desirable bell pepper-like and floral aroma notes, respectively.

14   **15**

Compounds Generated during Heating and Drying Processes

Numerous thermally-derived compounds are formed during the conversion of ripe Jalapeño peppers into dried chipotle peppers. Additional, some compounds may be lost during this process. For example, Luning et al. (*15*) reported that C_6 aldehydes responsible for "fresh" lettuce/grass/green notes

Table I. Aroma-Active Components of Dried Chipotle Peppers

No.	Compound	Odor Description[a]	RI[b]		Odor Intensity[c]			
			FFAP	DB5	A	B	C	Avg.
1	methylpropanal	malty, dark chocolate	821	<600	2	3	3	2.7
2	2-methylbutanal	bug, dark chocolate	912	655	5	4	4	4.3
3	3-methylbutanal	musty, chocolate	927	659	5	5	5	5.0
4	2,3-butanedione	buttery, creamy	981	594	1	1	3	1.7
5	ethyl 2-methylbutanoate	fruity, blueberry	1057	852	1	1	0	0.7
6	(Z)-4-heptenal	rancid, crabby	1230	962	2	2	3	2.3
7	1-octen-3-one[d]	mushroom	1301	979	5	3	3	3.7
8	(Z)-1,5-octadien-3-one[d]	metallic	1374	984	2	1	3	2.0
9	dimethyltrisulfide	sulfurous, cabbage	1387	971	2	5	3	3.3
10	2-isopropyl-3-methoxypyrazine[d]	earthy, soil	1431	1099	0	1	0	0.3
11	acetic acid	vinegar, pungent	1438	610	3	4	4	3.7
12	2-ethyl-3,5-dimethylpyrazine	potato, earthy	1445	1084	4	4	5	4.3
13	3-(methylthio)propanal (methional)	potato	1456	909	0	3	3	2.0
14	2-isobutyl-3-methoxypyrazine	bell pepper, earthy	1517	1175	5	5	4	4.7
15	(E)-2-nonenal[d]	hay, stale	1528	1161	3	0	2	1.7
16	linalool	floral, honeysuckle	1537	1103	3	4	3	3.3
17	2-methylpropanoic acid	Swiss cheese, bug	1555	- -	1	3	1	1.7
18	(E,Z)-2,6-nonadienal	cucumber	1587	1155	3	2	3	2.7
19	butanoic acid	cheesy, fecal	1618	824	3	4	3	3.3
20	phenylacetaldehyde	rosy, honey, plastic	1646	1046	3	2	2	2.3
21/ 22	2- and 3-methylbutanoic acid	sweaty, dried fruit	1659	878	5	4	5	4.7
23	2-methyl-(3-methyldithio)furan[e]	vitamin, bullion	1673	1173	2	1	4	2.3
24	unknown	hay, saffron	1730	- -	3	2	2	2.3
25	(E,E)-2,4-decadienal	fatty, fried, doughy	1817	1316	1	2	1	1.3
26	β-damascenone[d]	applesauce	1822	1387	4	4	4	4.0
27	2-methoxyphenol (guaiacol)	smoky	1861	1096	5	5	5	5.0
28	2-phenylethanol	rosy, wine	1915	1120	4	3	3	3.3
29	4-methylguaiacol	smoky, spicy	1944	1197	1	4	3	3.3

Table I. *Continued.*

No.	Compound	Odor Description[a]	RI[b]		Odor Intensity[c]			
			FFAP	DB5	A	B	C	Avg.
30	2-methylphenol (o-cresol)	phenol, inky	1999	1051	5	4	4	4.3
31	(E)-4,5-epoxy-(E)-2-decenal[d]	unripe, metallic	2005	1395	2	0	2	1.3
32	4-hydroxy-2,5-dimethyl-3(2H)-furanone (Furaneol)	burnt sugar, marshmallow	2022	1070	5	5	6	5.3
33	4-ethylguaiacol	smoky, cloves	2032	1285	2	1	3	2.0
34	octanoic acid	sweaty, waxy	2039	1282	4	4	4	4.0
35	4-methylphenol (p-cresol)	dung, horse stable	2077	1078	5	4	5	4.7
36	3-methylphenol (m-cresol)	bandaid, medicine	2084	- -	2	1	1	1.3
37	3-hydroxy-4,5-dimethyl-2(5H)-furanone (sotolon)	curry, spicy	2189	1115	5	4	3	4.0
38	2,6-dimethoxyphenol (syringol)	smoky, spicy	2264	1346	4	4	4	4.0
39	coumarin	herbaceous, woody	2465	1437	4	3	5	4.0
40	3-methylindole (skatole)	fecal, mothballs	2484	1402	4	3	4	3.7
41	phenylacetic acid	rosy, honey	2551	1047	4	3	5	4.0
42	3-methoxy-4-hydroxybenzaldehyde (vanillin)	vanilla	2568	1413	5	5	5	5.0

[a]Odor quality as perceived during GCO. [b]Retention indices calculated from GCO data, FFAP = Stabilwax™ DA (15 m x 0.32 mm i.d. x 0.5 μm film; Restek, Bellefonte, PA) column, DB5 = DB-5MS (15 m x 0.32 mm i.d. x 0.5 μm film; J&W Scientific, Folson, CA) column. [c]Odor intensity perceived at sniffing port (0 = not detected, 7 = very strong) determined on Stabilwax™DA column. A, B and C indicate different samples. [d]Compound tentatively identified based on comparison of odor property and retention indices with reference compound. [e]Compound tentatively identified based on comparison of odor property and retention indices with literature values.

disappeared or declined after hot-air drying of bell peppers. This observation may explain the aforementioned absence of these compounds in chipotle peppers.

The moist, and later dry heating conditions encountered during the production of chipotle peppers can lead to transformation of volatile compounds inherent to Jalapeño peppers or formation of new compounds from non-volatile precursors. Degradation of (E,Z)-2,6-nonadienal (no. **18**) via retro-aldol condensation (during heating/drying steps) is known to form (Z)-4-heptenal (no. **6**) (*17*). Methylpropanal (no. **1**), 2- and 3-methylbutanal (nos. **2** and **3**), 2,3-butanedione (no. **4**), dimethyltrisulfide (no. **9**), and acetic acid (no. **11**) were observed in dried bell peppers after rehydration (*18*). Hot-air drying of bell peppers increased the levels of 2-methylpropanoic acid (no. **17**), 2- and 3-methylbutanoic acids (no. **21/22**), 2-methylpropanal, and 2- and 3-methylbutanal (nos. **2** and **3**) as a result of the Strecker degradation of amino acid precursors (*15*). The above compounds also were observed as predominant odorants in two types of dried bell pepper powders by Zimmermann and Schieberle (*14*), who also reported methional (no. **13**), Furaneol (no. **32**), sotolon (no. **37**), and phenylacetic acid (no. **41**) as potent odorants. Butanoic acid (no. **19**) and skatole (no. **40**) were also identified, but at lower odor potency, in the dried bell peppers. These authors suggested that the two furanones (nos. **32** and **37**) could have been generated via hydrolysis of their glycoside precursors during drying. Similarly, β-damascenone (no. **26**) was probably thermally generated from a glycoside precursor as previously observed in the case of black tea (*19*).

Based on the above discussion it appears that numerous thermally-derived odorants may play an important role in chipotle aroma. Chemical structures of four of these compounds are shown below.

3

22

31

37

Compound Originating from the Smoking Process

The combustion of wood gives rise to numerous classes of volatile compounds. Among these the phenols, guaiacols and syringols are known to provide the characteristic phenolic and smoky notes to smoked foods (*20*). In the present study nine compounds (nos. **27, 29, 30, 31, 34, 36, 38, 39, 42**) were most likely derived during the wood smoking process. Among these, guaiacol (no. **27**) and syringol (no. **38**) provided intense smoky notes.

27   **38**

Conclusions

The characteristic aroma components of chipotle peppers are derived from the compounds inherent in the fresh Jalapeño peppers as well as the processing steps involved in their production. The compounds 2-isobutyl-3-methoxypyrazine and linalool impart fresh bell pepper-like and floral notes, respectively. Heating and dehydration steps lead to the formation of numerous compounds imparting mostly sweaty, malty/dark chocolate, and burnt sugar notes, such as 2- and 3-methylbutanoic acid, 3-methylbutanal, and Furaneol, respectively. Meanwhile, the smoking process provides the characteristic phenolic and smoky notes from phenol, guaiacols and syringol.

References

1. Huffman, V.L.; Schadle, E.R.; Villalon, B; Burns, E.E. *J. Food Sci.* **1978**, *43*, 1809-1811.
2. Chitwood, R.L.; Pangborn, R.M.; Jennings, W. *Food Chem.* **1983**, *11*, 201-216.
3. Acree, T. In *Flavor Measurement*; Ho, C.-T.; Manley, C.H. Eds.; Marcel Dekker, Inc.: New York, 1993; pp 77-94.
4. Blank, I. In *Techniques for Analyzing Food Aroma*; Marsili, R. Ed.; Marcel Dekker, Inc.: New York, 1997; pp 293-329.
5. Engel, W.; Bahr, W.; Schieberle, P. *Eur. Food Res. Technol.* **1999**, *209*, 237-241.

6. Ullrich, F.; Grosch, W. *Z. Lebensm.-Unters. Forsch.* **1987,** *184,* 277-282.
7. Schieberle, P.; Grosch, W. *Z. Lebensm. Unters. Forsch.* **1991,** *192,* 130-135.
8. Carunchia Whetstine, M.E.; Cadwallader, K.R.; Drake, M.A. 2005. *J. Agric. Food Chem.* **2005,** *53,* 3126-3132.
9. Karagul-Yuceer, Y.; Vlahovich, K.N.; Drake, M.A.; Cadwallader, K.R. *J. Agric. Food Chem.* **2003,** *51,* 6797-6801.
10. Van den Dool, H.; Kratz, P. *J. Chromatogr.* **1963,** *11,* 463-471.
11. Buttery, R.G.; Seifert, R.M.; Guadagni, D.G.; Ling, L.C. *J. Agric. Food Chem.* **1969,** *17,* 1322-1327.
12. Luning, P.A.; De Rijk, T.; Wichers, H.J.; Roozen, J.P. *J. Agric. Food Chem.* **1994,** *42,* 977-983.
13. Murray, K.E.; Whitfield, F.B. *J. Sci. Food Agric.* **1975,** *26,* 973-986.
14. Zimmermann, M.; Schieberle, P. *Eur. Food Res. Technol.* **2000,** *211,* 175-180.
15. Luning, P.A.; Carey, A.T.; Roozen, J.P.; Wichers, H.J. *J. Agric. Food Chem.* **1995,** *43,* 1493-1500.
16. Wu, C.-M.; Liou, S.-E. *J. Agric. Food Chem.* **1986,** *34,* 770-772.
17. Josephson, D.B.; Lindsay, R.C. *J. Am. Oil Chem. Soc.* **1987,** *64,* 132-138.
18. van Ruth, S.M.; Roozen, J.R. *Food Chem.* **1994,** *51,* 165-170.
19. Kumazawa, K.; Masuda, H. *J. Agric. Food Chem.* **2001,** *49,* 3304-3309.
20. Maga, J.A. *Food Rev. Int.* **1987,** *3,* 139-183.

Chapter 6

Analysis of *Piper auritum*: A Traditional Hispanic Herb

Brian G. McBurnett[1], Alfredo A. Chavira[1], A. Christina López[1], Jacklin Mosso[1], and Susan M. Collins[2]

[1]Department of Chemistry, University of the Incarnate Word, San Antonio, TX 78209
[2]Air Force Institute for Operational Health, Brooks City-Base, San Antonio, TX 78235

Volatile and semivolatile compounds in the leaves of *Piper auritum* were extracted and identified. The volatile chemical components of this herb were analyzed by static headspace analysis/gas chromatography/mass spectrometry (SHA/GC/MS). The semivolatile constituents of the leaves were isolated using both a hot water and Soxhlet extraction and characterized using GC/MS. The major component of *Piper auritum* was found to be safrole, which comprised over 60% of the relative concentration of all samples. Often used as a wrap for chicken or fish, a sample of chicken wrapped with *Piper auritum* showed migration of safrole into the interior of the meat.

Piper auritum is a small shrub native to the Central American tropics and a member of the diverse *Piperaceau* (Pepper) family (*1*). Depending on culture and the region where it is grown, *Piper auritum* is more commonly referred as hoja santa, acoyo, hoja del la estrella, acoyo, or momo in Mexico and Central America. In the United States, this herb is commonly referred as Mexican pepperleaf, or root beer plant (*2*).

Piper auritum is noted for its large ear-shaped leaves, which can grow up to 30 cm long as seen in Figure 1. The leaves possess a strong and distinctive aroma and as suggested in the name "root beer plant," its smell is similar to that of sassafras but also includes hints of anise and pepper. It has long been utilized for its medicinal and culinary properties (*3*).

Of the wide variety of medicinal applications of this herb, a majority of these remedies are focused on respiratory affliction. Typical treatment involves preparation of an infusion tea for treatment of bronchitis and throat irritations (*4*). *Piper auritum* has a long history of culinary applications. First used by the Aztecs in the preparation of ceremonial chocolate, it continues to be popular throughout Central America and in particular in the state of Veracruz, Mexico (*5*). *Piper auritum* is considered an essential ingredient in the preparation of *mole verde*, one of the famous "seven sauces of Oaxaca" (*6*). The leaves are also often used to wrap around a stuffing to be grilled or steamed (*3*). The most common dishes of this type are *pollo* (chicken) *o pescado* (fish) *en hoja santa.*

Our research focused on the determination of the chemical components of *Piper auritum*. In order to address the contribution of this herb to Hispanic cuisine, the focal points of this project include:

- Aroma (Analysis of volatile chemicals)
- Tea infusion (Hot water extraction)
- Use as a herb (Soxhlet extraction)
- Use as a wrap for chicken (Migration of chemical species)

As aroma is a significant component of flavor, our initial objective was to investigate the volatile chemicals of *Piper aurtium*. A hot water extraction was also prepared in order to emulate conditions similar to that of a typical tea made with the fresh leaf. Isolation of semi-volatile chemicals, as those found in sauces, was accomplished with a Soxhlet extraction and this method was also used to determine the contribution of *Piper auritum* to a grilled chicken sample wrapped in the leaf.

Materials and Methods

Materials

Freshly harvested *Piper auritum* leaves and grilled chicken samples were donated by Liberty Bar Restaurant in San Antonio, Texas. Samples were frozen in liquid nitrogen and crushed using a mortar and pestle.

Figure 1: Leaf of Piper auritum

Analysis of Volatile Chemicals

A 2.0 g sample of crushed *Piper auritum* was placed in a 22 mL headspace vial. A Tekmar 7000 autosampler provided the headspace sample to a Varian 3900 GC/Ion Trap MS. A 1 mL volume of the headspace sample was injected. The volatile components were analyzed with a Restek RTX-624 column (60 m x 0.32 mm x 1.8 μm film). The oven program started at 35°C for 2 min and ramped at 4°C/min to 50°C and then ramped at 10°C/min to 220°C and held at this final temperature for 5 min. The flow rate of the He carrier gas was 1.2 mL/min.

Analysis of Semivolatile Chemicals

Hot Water Extraction

A 5.0 g of sample of crushed *Piper auritum* was mixed with 200 mL of boiling water and left to infuse for 5 min. The leaves were removed by filtration and the resulting tea was allowed to cool to room temperature. The tea was extracted using a liquid/liquid separatory funnel technique. The tea was extracted with 250 mL CH_2Cl_2 and the resulting infusion was extracted with 400 mL of CH_2Cl_2 and separated into 3 fractions. In order to obtain a majority of the compounds in the sample, an acid and base extraction was performed. Ten molar sulfuric acid (H_2SO_4) and 10 M sodium hydroxide (NaOH) were used to adjust the pH level of the tea samples at a 1:1 ratio. The tea was first extracted at a normal pH using 250 mL of methylene chloride (CH_2Cl_2). The aqueous layer was acidified to a pH of 2 and extracted with an additional 250 mL of CH_2Cl_2 The pH of the sample was raised to 10 and extracted a final time with 250 mL CH_2Cl_2. All three methylene chloride extracts were combined and dried by passing it through a column containing 10 cm of anhydrous sodium sulfate. This extract was concentrated to 1 mL using a water bath and a gentle stream of nitrogen. Analysis of the sample was completed using HP 6890 GC with a 5973 MSD. Separation was achieved with a HP-5MS column: 30m x 0.25mm x 0.25 μm. The column was held at 35°C for 2 min then ramped at 35°C/min to 130°C and increased at 12°C/min to a final temperature of 310°C held for 5.43 min.

Soxhlet Extraction

The soxhlet extraction method was applied to isolate and concentrate semi volatile organics from the solid leaf sample. The analytical batch consisted of one sample and three controls – 10 g of crushed *Piper auritum,* a control blank, and a spiked blank along with a spiked blank duplicate. A 1.0 mL spiking solution of safrole was added to the spiked blank and spiked blank duplicate to measure efficiency of the procedure. Each sample was combined with sodium sulfate (Na_2SO_4) to remove residual moisture, and refluxed with 300 mL of methylene chloride for 12 hr. The extract was dried by passing it through a column containing anhydrous Na_2SO_4. A water bath with a gentle stream of nitrogen was used to concentrate the extract to 1.0 mL. Semivolatile components were analyzed using a HP 6890 GC with a 5973 MSD and a HP-5MS column. Ramp temperatures were identical to that of the hot water extraction previously described.

Migration of Chemical Species

The soxhlet extraction method was also used in the determination of the chemical components transferred while cooking with *Piper auritum*. Two chicken samples were grilled with one of the samples wrapped in a leaf of *Piper auritum*. A 10 g sample of the interior meat was dissected, frozen with liquid nitrogen, and crushed using a mortar and pestle. The samples were processed using the soxhlet extraction method.

Results and Discussion

Volatile components identified by headspace GC/MS are listed in Table I. The primary constituent in the leaves of *Piper auritum* is safrole at 70.9%, and was confirmed using a safrole standard. Remaining volatile compounds, as reported in Table I, were tentatively identified by comparison of their mass spectra and retention times to NIST library searches.

Extraction of the prepared tea sample also showed safrole to be the major component with a concentration over 70%. Minor components of the tea sample, reported in Table II, were tentatively identified by comparison of their mass spectra and retention times to NIST library searches.

Soxhlet extraction of the leaves of *Piper auritum* give a safrole concentration of 60.8% as shown in Table III. Depending on the technique utilized, safrole concentration of all extracts ranged from 60.8 to 75.7%. This corresponds closely to steam distillation of the leaves of *Piper auritum* with the concentration of safrole ranging from 65–77% (*7-10*). Minor components of the soxhlet extraction were identified by their retention times and masses. It is interesting to note that a majority of the compounds isolated in the leaves are terpene derivatives in contrast to the alkaloid compounds found in the root of the plant (*11, 12*).

Table I. Relative Percentages from FID of the Volatile Constituents Identified in the Leaves of *Piper auritum*

Component	Retention time (min)	Percentage
Safrole	16.51	70.9
Isosafrole	16.92	0.4
2-Hexenal	18.60	5.2
α-Phellandrene	19.15	0.7
α-Pinene	20.39	3.1
β-Pinene	20.65	3.7
β-Terpinene	21.21	3.7
γ-Terpinene	21.62	4.4
Limonene	23.25	6.1

Pollo en hoja santa (chicken wrapped in *Piper auritum* as shown in Figure 2) was prepared in order to determine the contribution of the leaf to the flavor of the chicken sample. Compared to control, there is an appreciable concentration of safrole in the interior of the meat at 47 ppm. Even if the leaf is not ingested as a part of this dish, there is transfer of its constituents in the process of cooking.

Safrole is responsible for the majority of the flavor found in *Piper auritum*, but has been found to be a potent hepatotoxin and hepatocancerogen in rats. This chemical is prohibited from direct addition to food by the Food and Drug Administration and is "reasonably anticipated to be a human carcinogen" (*13*). Although the concentration of safrole in cuisine cooked with *Piper auritum* is relatively low, the FDA warning should be taken into consideration when eating dishes prepared with the leaf.

Conclusion

In recent years, there has been an increase of interest in the use of natural herbs and their role in cuisine. For *Piper auriutm*, safrole was the major constituent found in all extracts and, although not directly ingested, did contribute to the chemical profile of the grilled chicken sample. Concentration of safrole varied according the extraction method and it is important to consider the preparation process when analyzing its contribution in cuisine.

Figure 2: Pollo en hoja santa, grilled chicken wrapped in Piper auritum

Table II. Relative Percentages of the Constituents in the Hot Water
Extraction of the Leaves of *Piper auritum*

Component	Retention time (min)	Percentage
4-Pentenal	3.94	0.4
4-Hydroxy-4-methyl-2-pentanone	4.29	0.7
2-Hexenal	4.35	1.1
Linaloyl oxide	5.13	0.2
(E)-3-Hexenoic acid	5.41	1.4
β-Cymene	5.43	1.4
α-Ocimene	5.51	0.7
(Z)-β-Ocimene	5.59	0.4
α-Phellandrene	5.68	1.3
cis-Linaloloxide	5.79	0.6
trans-Linaloloxide	5.89	0.6
Linalool	5.94	0.5
Myrcenol	6.07	0.3
1-Terpineol	6.12	0.6
β-Pinene	6.22	0.4
Myrcene	6.29	0.4
Terpinen-4-ol	6.55	0.8
α-Terpineol	6.69	4.9
2,6-Dimethyl-7-octene-2,6-diol	6.93	1.8
Safrole	7.49	75.7
Eugenol	7.96	0.7
Methyl eugenol	8.28	0.5
δ-Cadinene	9.42	0.5
Methoxyeugenol	10.12	3.0
3-Hydroxy-β-damascone	10.30	1.1

Acknowledgments

The authors would like to thank George Lee, Suk Bin Kong, Felix Chavez, and Chris Austin for technical assistance. We would also like to thank the Liberty Bar Restaurant in San Antonio, Texas for donation of plant and preparation of chicken samples.

Table III. Relative Percentages of the Constituents in the Soxhlet Extraction of the Leaves of *Piper auritum*

Component	Retention time (min)	Percentage
α-Pinene	4.90	0.7
β-Pinene	5.16	1.4
α-Phellandrene	5.18	0.3
Myrcene	5.23	1.0
β-Cymene	5.49	0.7
(Z)-β-Ocimene	5.52	1.2
(E)-β-Ocimene	5.59	0.9
γ-Terpinene	5.70	8.1
Terpinolene	5.90	1.8
Linalool	5.95	1.0
Terpinen-4-ol	6.56	1.0
Safrole	7.44	60.8
Eugenol	7.56	1.4
Copaene	8.19	1.4
α-Cubenene	8.29	0.8
β-Springene	8.60	3.2
Clovene	8.68	1.9
β-Caryophyllene	9.09	1.8
Germacrene D	9.15	3.8
Bicyclogermacrene	9.28	2.5
β-Bisabolene	9.48	1.3
(E)-Nerolidol	9.78	2.4

References

1. Jaramillo, M.A.; Manos, P.S. *Am. J. Bot.* **2001**, *88*, 706.
2. Gupta, M.P.; Monge, A.; Karikas, G.A.; Lopea de Cerain, A.; Solis, P.N.; de Leon, E.; Trujillo, M.; Suarez, O.; Wilson, F.; Montenegro, G.; Noriega, Y.; Snatana, A.I.; Correa, M.; Sanchez, C. *Int. J. Pharm.* **1996**, *34*, 19.
3. Castro, O.; Poveda, L.J. *Ing. Cienc. Quim.*, **1983**, *48*, 330.
4. Davidow, J. *Infusions of Healing: A Treasury of Mexican-American Herbal Remedies;* Simon & Schuster, Inc.: New York, 1999.
5. Bayless, R.; Bayless, D.G. *Authentic Mexican: Regional Cooking from the Heart of Mexico;* William Morrow & Co.: New York, 1987.
6. Gupta, M.P.; Arias, T.D.; Williams, N.H.; Tattje, D.H.E. *J. Nat. Prod.* **1985**, *48*, 330.

7. Castro, O.; Poveda, L.J. *Ing. Cienc. Quim.*, **1983,** *7,* 24.
8. Ciccio, F.J. *Ing. Cienc. Quim.*, **1995,** *15,* 39.
9. Pino, J.A.; Rosado, A.; Rodriquez, M.; Garcia, D. *J. Essent. Oil Res.* **1998,** *10,* 333.
10. Hansch, R.; Leuschke, A.; Gomez-Pompa, A. *Lloydia* **1975,** *38,* 529.
11. Nair, M.G.; Sommerville, J.; Burke, B.A. *Phytochemistry*, **1989,** *28,* 654.
12. IARC **1976,** *10,* 231.

Chapter 7

Phenolics, Flavonoids and Other Nutraceuticals in Mexican Wild Common Beans (*Phaseolus vulgaris*)

F. Guevara-Lara[1], L.G. Espinosa-Alonso[2], M.E. Valverde[2], A. Lygin[3], J. Widholm[3] and O. Paredes-López[2,*]

[1]Depto. de Química, Centro de Ciencias Básicas, Universidad Autónoma de Aguascalientes, Blvd. Universidad 940, Aguascalientes, Ags. 20100, México
[2]Depto. de Biotecnología y Bioquímica, Unidad Irapuato, CINVESTAV-IPN, Apdo. Postal 629, Irapuato, Gto. 36500, México
[3]Department of Crop Sciences, University of Illinois at Urbana-Champaign, 1201 West Gregory Drive 284 ERML, Urbana, IL 61801

Common beans (*Phaseolus vulgaris*) are a strategic food resource in many countries. They provide protein, vitamins and minerals, dietary fiber and other essential nutrients. Phytochemicals with nutraceutical activity in wild, weedy and cultivated genotypes include phenolic compounds, some of them providing color to the seeds. In view of their biodiversity, current efforts towards the identification of valuable wild and weedy genotypes will promote their conservation and rational utilization.

Dry common beans (*Phaseolus vulgaris* L.), also known as navy, pinto, red kidney or French beans, are widely consumed throughout the world. They are characterized by a high protein content and are an important source of energy, vitamins and minerals, and other essential nutrients. Common beans are especially valued as an inexpensive and nutritious food in many Latin-American and African countries [1, 2]. In Mexico, they are the second most important food (after corn). Besides its many nutrients, the common bean is rich in a variety of plant chemical compounds (also known as phytochemicals) with potential health benefits such as soluble and insoluble dietary fiber, and polyphenols [3]. Also, common beans contain a group of phytochemicals that have been historically regarded as antinutritional factors; these include trypsin inhibitors, hemagglutinins, phenolic compounds (tannins) and phytic acid. Numerous studies have led to the conclusion that phytic acid and tannins may bind proteins and some essential dietary minerals, thus making them unavailable or only partially available for absorption [4]. However, in view of abundant and recent evidence, the perspective on some of these compounds has been evolving to a more equilibrated opinion. The objective of the present work is to review the most current advances on the analysis of phenolic compounds and other nutraceuticals in common bean materials, a topic that has drawn much attention in recent years in view of the great potential benefits that those phytochemicals may bring to human health.

Phenolic compounds and their nutraceutical value

There is a great variability among common bean genetic materials. These materials include cultivated genotypes, those improved varieties grown for commercial purposes; wild materials, those that can be collected from the wilderness; and weedy genotypes, intermediate forms between cultivated and wild materials. Biodiversity is greater among wild and weedy materials, which makes them a very attractive resource for the search of novel agronomical and food quality traits.

Seed color and size are the two most importat quality characteristics for consumers [5, 6]. The seed color of beans is determined by the presence and concentration of phenolic compounds such as flavonoid glycosides, anthocyanins, and condensed tannins (proanthocyanidins) [7, 8].

Flavonoid compounds in common beans are reported to have biological activity in vitro and in vivo. These compounds have the properties of inhibiting autooxidation reactions and scavenging of free radicals [9]. Flavonoids extracted from beans, mainly anthocyanins and proanthocyanidins, have shown antioxidant [7, 10, 11] and antimutagenic [12-14] properties. Recently, red beans were identified as having one of the highest antioxidant capacities among over 100 common dietary fruits and vegetables analyzed [15]. Epidemiological studies suggest that the consumption of flavonoid-rich foods protects against human diseases associated with oxidative stress, like coronary heart disease and cancer. In vitro, flavonoids from several plant sources have shown free-radical scavenging activity and protection against oxidative stress, irradiation-induced cell damage, chemical-induced cellular transformation, and selective growth inhibitory activity of cancer cells [5].

The anthocyanins and anthocyanidins, aglycones of the anthocyanins, have aroused the interest of researchers, having shown, mostly in in vitro studies, beneficial effects for health. There are several reports on their anti-inflammatory, vasotonic, and anti-oxidant properties [10, 11]. They may, therefore, play an important role in the prevention of degenerative illnesses such as cancer, Alzheimer's disease or cardiovascular illnesses.

Phenolic acids also have antioxidant and other specific properties [16-18]. Protocatechuic acid has shown a chemopreventive effect on induced hepatocarcinogenesis in rats [19]. Also, several phenolic acids were capable of significantly altering (some positively and others negatively) the activity of phenolsulfotransferases (PSTs), important enzymes in the process of detoxification of xenobiotics and endogenous compounds in a human platelet assay. The overall effect of phenolic acids tested on the activity of PSTs was well corelated to their antioxidant activity [17].

Recently, four bean varieties (white kidney, red pinto, Swedish brown, and black kidney) were evaluated for their phenolic contents and antiradical activities. On the basis of the total phenolic content and antioxidant capacity, it was deduced that colored beans possess superior antioxidative activity compared with white beans [20, 21]. This conclusion agrees with those of two other recent reports. In one of them, 16 Mexican improved common bean cultivars were analyzed for antioxidant capacity, as well as for oligosaccharides, phytic acid, and trypsin inhibitory activity. The higher antioxidant capacities were associated to seed coat color [22], although more reserch is needed in this matter in view of reports of a high antioxidant activity associated to a high tannin content in white beans [7]. Also, there is not a clear relationship among total phenols, tannins and seed coat color in wild beans [23] or cultivated beans [24, 25]. The other work analyzed six bean cultivars grown in southern Manitoba and also reported higher antioxidant activities for those bean hull millstreams with higher total phenolic contents [26].

Flavonoids in common beans

The most widely distributed group of flavonoids in beans is proanthocyanidins (condensed tannins) [7, 24, 27). Proanthocyanidins have been detected in different varieties of common bean (9.4-37.8 mg catechin equivalents per g), mainly in the seed coat. [5, 7, 24, 27]; however, there relatively little comprehensive research on the proanthocyanidin profiles in beans.

The biological effects of flavonoids is dependent on the types of phytochemical constituents and the complexity of their structures, and the composition of the flavonoid mixtures, because it has been well established that complex mixtures of phytochemicals in fruits and vegetables may provide protective health benefits mainly through a combination of additive or sinergystic effects [28, 29]. In this vein, the characterization of polyphenolic compounds present in the seed coat of black Jamapa bean (a cultivated variety) was recently reported [5]. In this report, various fractionation methods were tested as to enable the recovery of polyphenolics in their naturally occurring forms. Direct silica gel fractionation of a 100% methanol extract allowed a more accurate identification of compounds, especially of the flavonols. Anthocyanins, flavanol monomers, and heterogeneous flavanol oligomers up to hexamers were detected. Also, myricetin glycoside and proanthocyanidin oligomers containing (epi)-gallocatechin were reported for the first time in black beans [5].

Another recent work reported that flavonoids, and not phenolic acids, were the main phenolic compounds in fractions from a white bean extract prepared using 80% (v/v) acetone. Interestingly, the authors reported that in these white bean fractions, the antioxidant activity did not correlate exactly with their content of total phenolic compounds [30].

The scarcity of knowledge on phenolic compounds in wild and weedy bean genetic materials has sparked much interest in recent years. Table 1 shows the summarized results of the analysis of 62 Mexican wild and weedy common bean collections (accessions), in regards to their total phenols, condensed tannins and anthocyanin contents [23, 31]. As a result of this work, several interesting wild and weedy genotypes have been identified, some of which are shown in Figure 1. This valuable information is being incorporated into common bean breeding programs, which will promote the conservation and rational utilization of unique and endangered germplasm resources.

Anthocyanins in common beans

The anthocyanins constitute one of the major groups of natural pigments and are responsible for many of the colors of fruits and vegetables as well as flowers. In recent years, several studies have been caried out to characterize the anthocyanin profiles of different natural products, among them the common bean, to use this type of phenolic compounds as an alternative to the synthetic colorants used in the food industry. From a technological point of view, one of the main advantages of the anthocyanins is their solubility in water, which facilitates their incorporation in foodstuffs [32].

The presence of anthocyanins has only been reported in black and blue-violet colored beans [5, 7]. Early studies focused on anthocyanin pigments in the seed coat of dry kidney beans [1, 33]. Four anthocyanins, namely malvidin 3-glucoside, petunidin 3-glucoside, delphinidin 3-glucoside, and delphinidin 3,5-diglucoside were extracted from black-violet beans [33]. Later, other researchers reported the isolation and characterization of anthocyanins from different common bean varieties [34, 35, 36]. Recent studies have dealth with antioxidant activity, germplasm characterization or separation, and structural elucidation of flavonol glycosides, isoflavones, and anthocyanins from common bean samples and their contribution to seed coat color [7, 10, 32, 37-40].

Table 1. Ranges of total phenols, condensed tannins and total anthocyanin content[a] from wild and weedy common beans from Mexico

Origin or cultivar	Type	N[b]	Total phenols mg GAE g^{-1}	Condensed tannins mg CAE g^{-1}	Total anthocyanin mg C3G g^{-1}
Chiapas	weedy	1	1.57	20.3	0.31
Chihuahua	wild	1	1.23	17.6	0.01
Durango	wild	10	1.05 – 1.64	12.1 – 27.2	0.03 – 0.47
Durango	weedy	3	1.05 – 2.11	12.9 – 35.7	0.03 – 0.40
Guerrero	wild	4	1.34 – 1.59	10.0 – 24.8	0.09 – 1.52
Jalisco	wild	11	0.91 – 1.55	9.5 – 21.7	0.02 – 0.51
Jalisco	weedy	3	1.01 – 1.45	10.9 – 15.7	0.04 – 1.74
Michoacan	wild	7	1.36 – 1.70	13.8 – 24.7	0.03 – 0.50
Michoacan	weedy	4	1.09 – 1.78	14.7 – 28.7	0.03 – 0.24
Morelos	wild	4	1.35 – 1.55	11.0 – 24.3	0.05 – 0.88
Morelos	weedy	1	1.30	14.3-	0.25
Nayarit	wild	1	1.39	24.6	0.18
Oaxaca	wild	10	1.23 – 1.90	12.8 – 24.6	0.04 – 0.51
Sinaloa	wild	1	1.08	18.1	0.18
Zacatecas	wild	1	1.07	11.1	0.43
Jamapa	cultivated	1	1.98	21.4	1.84
Pinto	cultivated	1	1.41	30.9	0.03

[a] Ref. [23, 31] Values are means of three determinations (fresh basis) per accession. Total phenols expressed as mg equivalents of gallic acid g^{-1}; condensed tannins expressed as mg equivalents of (+) catechin g^{-1}; total anthocyanin expressed as mg of cyanidin 3-glucoside g^{-1}. [b] N: Number of accessions analyzed.

In addition, a comparison has been recently carried out among methods such as rapid spectrophotometric methods for estimation of total phenolic content, and total flavonoid content with HPLC (DAD) and HPLC/MS using 12 bean samples collected from two different regions of Italy over a 3-year time period [37].

The aplication of improved methods of extraction and purification has allowed the characterization of anthocyanins and free anthocyanidins in beans [32]. HPLC-MS analyses have recently made it possible to identify seven anthocyanins, out of which cyanidin and pelargonidin mono- and diglucosides were outstanding. Also, some acytlated anthocyanins were reported, together with cyanidin, peonidin, and pelargonidin free aglycones (anthocyanidins) [32].

Another recent work reported the isolation of two anthocyanins from black common beans by semipreparative HPLC, followed by identification using LC-MS, spectroscopic UV-Vis and NMR methods. Their chemical structures were elucidated as delphinidin 3-glucoside and petunidin 3-glucoside [41].

82

Figure 1. Four Mexican wild and weedy accessions of Phaseolus vulgaris identified as high in phenolic acids, flavonoids, and anthocyanins. Source: References 23, 31.

Phenolic acids in common beans

The challenges asociated with the analysis of phenolic acids arise from their structural complexities, as these compounds may exist in multiple forms as free, esterified, glycosylated or polymerized [1, 42]. In addition, these compounds are not uniformly distributed in plants at tissue, cellular, and sub-cellular levels, and may coexist as complexes with proteins, carbohydrates, lipids or other plant components. Thus, the polarity of phenolic acids varies widely, and it is difficult to develop a uniform analysis procedure for an assay of all phenolic acids [1]. The first reports on the analysis of phenolic acids from common bean samples were carried out mainly to evaluate the impact of phenolic acids on the organoleptic characteristics of different varieties [43-45]. The authors used classical extraction procedures followed by base hydrolysis of bean extracts. No antioxidants were added during extraction and base hydrolysis to prevent the degradation of phenolic acids.

Recently, it has been observed that black beans contain significant amounts of phenolic acids. A systematic study was undertaken to separate, identify and quantify phenolic acids from 10 bean classes along with multiple varieties from three commonly cultivated and consumed classes (Pinto, Great Northern, and Black beans) in the United States [1]. All extractions and analyses were carried out by recently developed procedures in the presence of ascorbic acid and ethylenediaminetetraacetic acid. Sixteen phenolic acids were separated and quantified usind an HPLC/DAD procedure. Ferulic acid (the most abundant), p-coumaric acid and synapic acid were detected and quantified in all varieties. Caffeic acid was detected in measurable amounts in only two Black bean varieties. Total phenolic acid content among all samples varied between 19.1 and 48.3 mg/100 g of bean sample, with an average of 31.2 mg/100 g. Also, the effect of bean soaking and cooking on phenolic acid contents was evaluated. Over 83% of the total phenolic acids were retained in bean samples during the cooking process, and only 2% or less were detected in water extracts during overnight soaking [1].

A similarly recent work reports the analyses of several Mexican wild and cultivated common bean genetic materials regarding the flavonoid and phenolic acid contents. Also, the effect of cooking and germination on those phenolic compounds was evaluated (Table 2). In general, cooking caused a decrease in quercetin and kaempferol levels in all materials, while germination showed varied effects depending on the bean genetic material. Similar trends were observed for the phenolic acids analyzed [25].

Other nutraceuticals in common beans

In developing countries, cereal grains and some legumes are the primary and least expensive sources of calcium, iron and zinc; however, their levels in and intake from those sources does not satisfy the mineral requirements of the populations of those countries [4, 46]. It has been reported that the percentage of anemic subjects in developing nations (26%) is much higher than that observed in Europe (11%) and the US (8%); data revealed that anemia was predominantly caused by iron deficiency. [47, 48]. The same studies have found that 40% of the iron intake is derived from cereals and legumes. Recent reports indicate that iron deficiency is the most prevalent micronutrient problem in the world, affecting over 2 billion people globally, many of whom depend on different types of beans as their staple food [49].

Zinc is essential for normal growth, appetite and immune functionality, being an essential component of more than 100 enzymes involving digestion, metabolism and wound healing [50]. While iron deficiency has long been considered a major nutritional problem, zinc deficiency has only recently been recognized as a public health problem [3].

Table 2. Quercetin, kaempferol, and phenolic acid content[a] (µg/g, dry basis) in raw, cooked and 72-h germinated Mexican wild and cultivated common bean seeds.

Bean genotype	FMM38	N8025	G-12892	G-12906
Type	cultivated	cultivated	wild	wild
Seed coat color	beige-red	black	beige	black-brown
Weight of 100 seeds (g)	30.3	22.3	5.3	11.6
Quercetin				
Raw	6.9	9.7	11.3	17.9
Cooked	4.8	8.5	7.6	6.1
Germinated	16.5	50.4	15.1	13.3
Kaempferol				
Raw	16.1	19.2	23.1	37.0
Cooked	11.4	15.7	14.3	13.2
Germinated	19.7	20.7	11.8	10.5
HBA				
Raw	13.1	9.6	9.6	11.3
Cooked	7.6	7.5	7.3	8.6
Germinated	2.2	3.1	2.8	2.0
VA				
Raw	7.2	13.0	8.6	16.6
Cooked	4.5	9.8	Nd	12.1
Germinated	43.2	40.8	25.7	46.3
CMA				
Raw	6.2	6.8	5.6	5.3
Cooked	2.1	4.7	4.0	3.9
Germinated	Nd	15.5	Nd	Nd
FA				
Raw	20.8	36.0	24.6	28.4
Cooked	13.2	27.9	19.8	23.3
Germinated	25.5	41.9	26.3	22.7

[a]Ref. [25] Values are means of three determinations. [b]Not determined. HBA, p-hydroxybenzoic acid; VA, vanillic acid; CMA, p-coumaric acid; FA, ferulic acid.

The search for and utilization of useful genes from wild species for improvement of cultivated materials has been practiced with varying degrees of success in different crops. In order to improve quantitative traits in cultivated beans, the introduction of novel genetic variation through the introduction of exotic bean germplasm, such as wild and weedy common bean types, has been suggested [4, 51]. Owing to the relative ease with which wild beans can hybridize with cultivated forms, it is imperative that concerted efforts be made to collect these types rom egions not yet covered adequately in germplasm banks. In addition, systematic evaluation and utilization must be made in improvement programs.

Few efforts have been made to improve mineral content in cultivated common bean varieties by breeding programs. As a result, the contribution of genetic factors to mineral accumulation is poorly understood [52]. Few studies have investigated the mineral content as well as other nutritional and antinutritional components of wild and weedy common beans in order to gain knowledge of their genetic potential for the improvement of cultivated beans. Such information is an essential requirement before hereditability and/or environmental studies can be conducted.

As a response to such need of information, 70 accessions of wild and weedy common bean from different sites in two Mexican states (Jalisco and Durango) were analyzed for protein, total calcium, iron and zinc [4]. In addition, protein digestibility, essential amino acid profiles, tannins, phytic acid, and extractable iron were determined in selected accessions. The phytate/zinc and the phytate × (calcium/zinc) molar ratios were also determined as predictors of zinc bioavailability. For compartive purposes, two cultivated common bean varieties were included in the study. The wild and weedy beans contained more protein and similar protein digestibility as compared with the cultivated materials. The contents of sulfur amino acids were low in all samples; additionally, beans from Jalisco had higher contents of sulfur amino acids than the cultivated genotypes. Beans from Durango showed higher leucine, valine and aromatic amino acids than cultivated beans. Some wild and weedy beans from Jalisco and Durango showed high contents of calcium (7470 mg kg^{-1}), iron (280 mg kg^{-1}), and zinc (33.1 mg kg^{-1}). The phytic acid × (calcium/zinc) molar ratios of some wild and weedy beans were similar to those of cultivated beans. Amounts of extractable iron were in the order of 26-74%. Useful wild and weedy bean accessions were identified and may well be utilized to conduct hereditability studies in plant breeding programs to improve the nutritional characteristics of cultivated common beans [4].

A more recent work reported the identification of putative quantitative trait loci (QTL) associated with seed mass, calcium, iron, zinc, and tannin content in bean seeds. A total of 57 AFLP markers were distributed among five linkage groups with a coverage of 497 centiMorgans [53]. Five putative QTL were significantly associated with seed mass, two with calcium, two with iron, one with zinc and four with tannin content in the seed. These QTL explained approximately 42, 25, 25, 15 and 42% of the phenotypic variance, respectively. Due to known environmental effects on most nutritional traits, QTL could be used to screen segregating populations that include wild genotypes, wild populations, and ancestral landraces from the region where outstanding wild populations are identified [53].

Conclusions

Common beans are being widely recognized as a most important source of nutrients and health promoting messages, both in developing and developed countries. In recent years a wealth of information has been gathering on the nutrient and nutraceutical composition of this important legume, in particular from wild and weedy genetic materials. As new collections of these rare and valuable materials are being gathered, particularly from regions regarded as centers of origin of the species, a treasure of novel genes is being saved from genetic erosion, a phenomenon that has been severely affecting this species. Thanks to this appropriation of genotypes and information, the agronomical, food quality, nutritional and nutraceutical characterization of these materials is being actively incorporated into plant breeding programs aimed at improving the quality of current commercial cultivars. This is enabling the conservation and rational utilization of wild and weedy common beans, which will ultimately be reflected in a crop with improved acceptance by the consumer.

References

1. Luthria, D.L.; Pastor-Corrales, M.A. *J. Food Comp. Anal.* **2006**, 19, 205-211.
2. Shellie-Dessert, K.C.; Bliss, F.A. In *Common Beans: Research for Crop Improvement*. A. Schoonoven and O. van-Voysest (Eds.), CAB International, Wallingford, U.K., **1991**; pp 649-673.
3. Guzmán-Maldonado, S.H.; Paredes-López, O. In Functional Foods. Biochemical and Processing Aspects. G. Mazza (Ed.), Technomic, Lancaster, PA, **1998**; pp 239-328.
4. Guzmán-Maldonado, S.H.; Acosta-Gallegos, J.; Paredes-López, O. *J. Sci. Food Agric.* **2000**, 80, 1874-1881.
5. Aparicio-Fernández, X.; Yousef, G.G.; Loarca-Piña, G.; de Mejía, E.; Lila, M.A. *J. Agric. Food Chem.* **2005**, 53, 4615-4622.
6. Castellanos, J.Z.; Guzman-Maldonado, S.H.; Jiménez, A.; Mejia, C.; Muñoz-Ramos, J.J.; Acosta–Gallegos, J.A.; Hoyos, G.; Lopez-Salinas, E.; Gonzalez-Eguiarte, D.; Salinas-Perez, R.; Gonzalez-Acuña, J.; Muñoz-

Villalobos, J.A.; Fernández-Hernández, P.; Cazares, B. *Arch. Latinoamer. Nutr.* **1997**, 47, 163-167.
7. Beninger, C.W.; Hosfield, G.L. *J. Agric. Food Chem.* **2003**, 51, 7879-7883.
8. Beninger, C.W.; Hosfield, G.L.; Bassett, M.J. *J. Amer. Soc. Hort. Sci.* **1999**, 124, 514-518.
9. Burda, S.; Oleszek, W. *J. Agric. Food Chem.* **2001**, 49, 2774-2779.
10. Cardador-Martínez, A.; Loarca-Piña, G.; Oomah, B.D. *J. Agric. Food Chem.* **2002**, 50, 6975-6980.
11. Tsuda, T.; Ohshima, K.; Kawakishi, S.; Osawa, T. *J. Agric. Food Chem.* **1994**, 42, 248-251.
12. Cardador-Martínez, A.; Castaño-Tostado, E.; Loarca-Piña, G. *Food Add. Contam.* **2002**, 19, 62-69.
13. Aparicio-Fernández, X.; Manzo-Bonilla, L.; Loarca-Piña, G.F. *J. Food Sci.* **2005**, 70, S73-S78.
14. González de Mejía, E.; Castaño-Tostado, E.; Loarca-Piña, G. *Mut. Res.* **1999**, 441, 1-9.
15. Wu, X.L.; Beecher, G.R.; Holden, J.M.; Haytowitz, D.B.; Gebhardt, S.E.; Prior, R.L. *J. Agric. Food Chem.* **2004**, 52, 4026-4037.
16. Pietta, P. *J. Nat. Prod.* **2000**, 63, 1037–1042.
17. Yeh, C.; Yen, G. *J. Agric. Food Chem.* **2003**, 51, 1474 – 1479.
18. Murphy, P. A.; Hendrich, S. *Adv. Food Nutr. Res.* **2002**, 44, 195 – 246.
19. Tanaka, T.; Kojima, T.; Kawamori, T.; Yoshimi, N.; Mori, H. *Cancer Res.* **1993**, 53, 2775-2779.
20. Madhujith, T.; Shahidi, F. *J. Food Sci.* **2005**, 70, S85-S90.
21. Madhujith, T.; Naczk, M.; Shahidi, F. *J. Food Lipids* **2004**, 11, 220-233.
22. Iniestra-González, J.J.; Ibarra-Pérez, F.J.; Gallegos-Infante, J.A.; Rocha-Guzmán, N.E.; González Laredo, R.F. *Agrociencia* **2005**, 39, 603-610.
23. Espinosa-Alonso, L.G.; Lygin, A.; Widholm, J.M.; Valverde, M.E.; Paredes-López, O. *J. Agric. Food Chem.* **2006**. Submitted for publication.
24. Guzmán-Maldonado, S.H.; Castellanos, J.; González de Mejía, E. *Food Chem.* **1996**, 55, 333-335.
25. Díaz-Batalla, L.; Widholm, J.M.; Fahey, G.C.; Castaño-Tostado, E.; Paredes-López, O. *J. Agric. Food Chem.* **2006**. In press.
26. Oomah, B.D.; Cardador-Martínez, A.; Loarca-Piña, G. *J. Sci. Food Agric.* **2005**, 85, 935-942.
27. De Mejía, E.G.; Guzmán-Maldonado, S.H.; Acosta-Gallegos, J.A.; Reynoso-Camacho, R.; Ramírez-Rodríguez, E.; Pons-Hernández, J.L., González-Chavira, M.M.; Castellanos, J.Z.; Kelly, J.D. *J. Agric. Food Chem.* **2003**, 51, 5962-5966.
28. Seeram, N.P.; Adams, L.S.; Hardy, M.L.; Heber, D. *J. Agric. Food Chem.* **2004**, 52, 2512-2517.
29. Shafiee, M.; Carbonneau, M.; Huart, J.; Descomps, B.; Leger, C. *J. Med. Food* **2002**, 5, 69-78.
30. Karamac, M.; Amarowicz, R. *J. Food Lipids* **2004**, 11, 165-177.

88

31. Espinosa-Alonso, L.G.; Valverde, M.E.; Guevara-Lara, F.; Paredes-López, O.; Lygin, A.; Widholm, J. Chemistry and Flavor of Hispanic Foods; 229th ACS National Meeting, San Diego, CA, 2005, AGFD 094, p. TECH-3.

32. Macz-Pop, G.A.; Rivas-Gonzalo, J.C.; Pérez-Alonso, J.J.; González-Paramás, A.M. Food Chem. 2006, 94, 448-456.

33. Feenstra, W.J. Mededelingen Landbouwhogeschool Wageningen, 1960, 60, 1-53.

34. Okita, C.; Kazuko, S.; Kazuko, Y.; Hamaguchi, Y. Eiko To Shokuryo, 1972, 25, 427-430.

35. Stanton, W.R.; Francis, B.J.; Nature 1966, 211, 970-971.

36. Takeoka, G.R.; Dao, L.T.; Full, G.H.; Wong, R.Y.; Harden, L.A.; Edwards, R.H.; Berrios, J.D.J. J. Agric. Food Chem. 1997, 45, 3395-3400.

37. Heimler, D.; Vignolini, P.; Dini, M.G.; Romani, A. J. Agric. Food Chem. 2005, 53, 3053-3056.

38. Choung, M.-G.; Choi, B.-R.; An, Y.-N.; Chu, Y.-H.; Cho, Y.-S. J. Agric. Food Chem. 2003, 51, 7040-7043.

39. Romani, A., Vignolini, P., Galardi, C.; Mulinacci, N.; Benedettelli, S.; Heimler, D. J. Agric. Food Chem. 2004, 52, 3838-3842.

40. Tsuda, T.; Shiga, K.; Ohshima, K.; Kawakishi, S.; Osawa, T. Biochem. Pharmacol. 1996, 52, 1033-1039.

41. Choung, M.G. Food Sci. Biotechnol. 2005, 14, 672-675.

42. Robbins, R.J. J. Agric. Food Chem. 2003, 51, 2866-2887.

43. Drumm, T.D.; Gray, J.I.; Hosfield, G.L. J. Sci. Food Agric. 1990, 51, 285-297.

44. Sosulski, F.W.; Dabrowski, K.J. J. Agric. Food Chem. 1984, 32, 131-133.

45. Schmidtlein, H.; Hermann, K.Z. Z. Lebensm. Unters-Forsch. 1975, 159, 213-218.

46. Reyes-Moreno, C.; Paredes-López, O. Crit. Rev. Food Sci. Nutr. 1993, 33, 227-286.

47. Rosado, J.L.; López, P.; Morales, M.; Muñoz, E.; Allen, L.H. British J. Nutr. 1992, 68, 45-58.

48. Barclay, D.V.; Heredia, L.; Gil-Ramos, J.; Montalvo, M.M.; Lozano, R.; Mena, M.; Dirren, H. Arch. Latinoam. Nutr. 1996, 46, 122-127.

49. DellaPena, D. Science 1999, 285, 375-379.

50. Stauffer, J.E. Cereal Foods World 1999, 44, 115-117.

51. Schneider, K.; Kelly, J.D. Michigan Dry Bean Digest 1999, 23, 15-20.

52. Moraghan, J.T.; Grafton, K. J. Sci. Food Agric. 1997, 74, 251-256.

53. Guzmán-Maldonado, S.H.; Martínez, O.; Acosta-Gallegos, J.A.; Guevara-Lara, F.; Paredes-López, O. Crop Sci. 2003, 43, 1029-1035.

Chapter 8

Chemical Parameters and Biological Activity of Phenolic Compounds in *Phaseolus vulgaris* and *Phaseolus coccineus* Beans

Guadalupe Loarca-Piña[1,3], H. S. Guzmán-Maldonado[2], J. Acosta-Gallegos[2], A. Álvarez-Muñoz[1], and S. García-Delgado[1]

[1]PROPAC, Research and Graduate Studies in Food Science, School of Chemistry, University of Querétaro, Qro., Mexico 76010
[2]Biotechnology Unit and Common Bean Program, Experimental Station El Bajío, National Research Institute for Forestry, Agriculture and Livestock (INIFAP), Carretera Celaya-San Miguel de Allende, Guanajuato, Mexico
[3]Current address: Departamento de Investigación y Posgrado en Alimentos, Facultad de Química, Centro Universitario, Cerro de las Campanas S/N, Querétaro, Qro., Mexico 76010

The potential health benefits of consuming beans as nutraceuticals have largely been overlooked. The seed coat of dry beans is rich in phenolic compounds, which are effective antimutagens and anticarcinogens. The objective of this study was to evaluate the effects of phenolic compound concentration on the antioxidant and antimutagenic activities of polyphenols from the seed coat of *Phaseolus vulgaris* and *Phaseolus coccineus* black seeded cultivars. The antioxidant potential was evaluated *in vitro*. The antimutagenic activity of phenolic compounds against aflatoxin B_1 was tested using a microsuspension assay. The methanolic extracts from the seed coat exhibited antioxidant activity that correlated with phenolic content. *P. coccineus* had the highest content of phenolic compounds (925.55 ± 73 mg eq. (+) catechin/g). *P. vulgaris* showed higher values of anthocyanins than *P. coccineus* (60.74 ± 4.25 and 8.64 ± 0.43 mg eq. (+) cyanidin-

3-glucoside/kg, respectively). The antimutagenicity of phenolic compounds from *P. coccineus* cv 'Ayocote Negro' was higher than that of *P. vulgaris* cv 'Negro Jamapa.' Due to the antimutagenic activity of their chemical constituents, these beans could be used as nutraceuticals or as ingredients in functional foods.

Introduction

The common bean (*Phaseolus vulgaris*) is the most important food legume and is a major source of dietary protein in many Latin American countries. In Mexico, the estimated per capita intake of beans is 22 kg/yr (*1*). The consumption of dry common beans has been associated with a reduction in chronic diseases such as cancer, diabetes, obesity and cardiovascular disease (*2-4*). Common bean seeds contain many phytochemicals including polyphenols which may be the responsible for these physiological effects. Our previous studies (*5-7*) showed that the seed coat of beans contains a large amount of phenolic compunds which display antioxidant and antimutagenic activities.

Phaseolus coccineus is another legume species of the genus Phaseolus, that grows in Mexico, Central America, South America, Asia, and in the United Kingdom (*8*). Although the chemical composition of this particular bean is similar to that of the common bean (*9, 10*) and they are phylogenetically related (*11*), its potential value as a nutraceutical has not been studied.

The objective of this study was to measure the phenolic concentration from the seed coat of *Phaseolus vulgaris* and *Phaseolus coccineus* black seeded cultivars and to determine their antioxidant and antimutagenic activities.

Materials and Methods

Beans

Dry beans (*Phaseolus vulgaris* Negro Jamapa and *Phaseolus coccineus* Ayocote Negro cultivar) were kindly provided by The Bean Program of the Instituto Nacional de Investigaciones Forestales Agricolas y Pecuarias campus El Bajío (INIFAP, Mexico). All samples were stored in the dark at 4°C until processed and analyzed.

Chemicals

Aflatoxin B_1 (AFB$_1$), (+)-catechin, vanillin, butylated hydroxytoluene (BHT), 1,1-diphenyl-2-picrylhydrazyl (DPPH), 6-hydroxy-2,5,7,8-tetramethylchroman-2-carboxylic acid (Trolox), enzymes, and dimethyl sulfoxide (DMSO) were from Sigma Chemical Co. (St. Louis, MO). Stock solution of AFB$_1$ was prepared in DMSO at 0.1 mg/mL.

Extraction of Total Polyphenolic Compounds and Total Anthocyanins

Extraction of total polyphenolic compounds and total anthocyanins of seed coats followed the method of Cardador-Martínez et al. (6). Lyophilized samples (200 mg) were placed in a 50-mL flask and mixed with 10 mL methanol. The flask was protected from light and was shaken for 24 hr at 25°C. After shaking, the samples were centrifuged at 2500 rpm for 10 min and analyzed.

Quantification of Total Polyphenolic Compounds and Total Anthocyanins

Total polyphenolic compounds, expressed as mg of (+)-catechin equivalents per gram of sample, were analyzed according to Deshpande and Cheryan (12, 13). Briefly, 5 mL of vanillin reagent (0.5% vanillin, 4% HCl in methanol) were added to 1 mL of methanolic extract solution. The reaction was carried out at 30°C for 20 min. Total polyphenolic compounds were quantified by spectrophotometry using (+)-catechin (up to 0.2 mg/mL) as a reference standard. To correct for potential interferences from natural pigments in beans, a blank sample was prepared by subjecting the original extract to the same conditions of reaction but without adding the vanillin reagent.

Total anthocyanins are expressed as mg of cyanidin-3-glucoside equivalents per gram of sample, following Al-Saikan et al. (14). Briefly, 30 mg of lyophilized sample was added to a 50-mL flask and mixed with 12.5 mL acidified ethanol (ethanol:1N HCl, 85:15, v/v), the mixture was shaken for 30 min at 25°C and then the pH was adjusted to 1. After the incubation, the samples were centrifuged at 4500 rpm for 15 min, and the supernatant was poured into a 25 mL volumetric flask. The volume was then adjusted with acidified ethanol. Anthocyanin absorption was measured by spectrophotometry and quantified according to the equation in Al-Saikan et al. (14).

Theorical Polyphenols Contents

The theoretical amount of polyphenols [TAP (condensed tannins or anthocyanins)] in whole raw bean seed was calculated from the concentration in the seed coat by TAP = (SCP) x (SC)/100, where SCP is the seed coat phenolic content (mg phenolic extract/g) and SC is the seed coat percentage of the whole beans (*15*).

Protein and *in vitro* Protein Digestibility

Bean seeds were milled into flour (mesh 60). Protein contents were determined by the Kjeldahl method (*16*); conversion factor was N×6.25. *In vitro* protein digestibility of cooked bean flours was estimated using the equation Y = 234.84 - 22.56X, where Y is *in vitro* protein digestibility (%) and X is the pH of the protein sample suspension after 20 min of proteolitic multi-enzyme digestion (*16*).

Water Absorption

The procedure described by Castellanos et al. (*17*) was followed to ascertain bean water absorption. Whole seed samples (10.0 ± 1.0 g) were soaked in 50 mL of distilled water at 25°C. At predetermined time intervals, the beans were removed from the soaking water, drained, surface-dried with filter paper and reweighed. The water absorption was determined and expressed as the increase (%, w/w) in weight of dry beans. Bean samples (25 seeds) were cooked in 100 mL of distilled water at 94°C.

Cooking Time

A Mattson bean cooker with minor modifications was used to evaluate 25 seeds at a time. Cooking time is the mean time, over three replications, when 19 of the beans (75%) were cooked, as indicated by plungers dropping and penetrating individual beans. The 75% cooked point corresponds to the sensorially preferred degree of cooking (*18*). Cooking time for each cultivar was the time averaged over two growing replications (*17*).

Mutagenicity and Antimutagenicity Assays

The Kado microsuspension assay, which is a simple and sensitive modification of the Ames test, was used throughout (19, 20). Tester strains TA98 and TA100 were kindly provided by Dr. B.N. Ames, Berkeley, CA. For the microsuspension procedure, bacteria were grown overnight in Oxoid Nutrient Broth No.2 (Oxoid, UK) to ~ 1-2 x 19^9 cells/ml. The cells were collected by centrifugation (4500 xg, 10 min, 4°C) and re-suspended in ice-cold phosphate buffered saline (0.15 M PBS, pH 7.4). The S9 (metabolic enzymes) and S9 mix (enzyme + cofactors) were prepared according to Ames et al. (21). The S9 from Aroclor 1254 pre-treated male Sprague-Dawley rats was obtained from MOLTOX, Inc. (Boone, NC) and protein concentration was adjusted to 300 µg protein/mL according to the concentration given by MOLTOX, Inc. The following ingredients were added to 12 x 75 mm sterile glass culture tubes kept on ice: 0.1 mL S9 mix, 0.1 mL concentrated bacteria in PBS (10^{10} cells/mL PBS), 0.01 mL AFB$_1$ solution (0.125, 0.25, 0.5 and 1 µg/tube), or 0.01 mL of phenolic extract (PE) from bean seed (10, 50, 100, 200 and 400 µg-equivalent (+)-catechin/tube) or 0.005 mL PE from bean seed (10, 50, 100, 200 and 400 µg-equivalent (+)-catechin/tube) + 0.005 mL of AFB$_1$ (0.5 µg/tube) for TA100 and TA98. The mixture was incubated in the dark at 37°C with rapid shaking. After 90 min, the tubes were placed in an ice bath. Tubes were removed one at the time, and 2 mL of molten-top agar containing 90 nmol of histidine and biotin was added (21). The combined solutions were vortex-mixed and poured onto minimal glucose plates. Plates were incubated at 37°C in the dark for 48 hr after which the numbers of revertant colonies were counted using a colony counter (AccuCount 1000, BioLogics Inc.). Strain markers and bacterial survival were routinely monitored in each experiment. Samples were tested in triplicate for each independent experiment that was performed.

β-Carotene Bleaching Method

The antioxidant activity of the phenolic extracts was evaluated according to the β-carotene-linoleate model system (22) based on the procedures of Marco (23) and Velioglu et al. (24). Aliquots (20 µL) of each extract or BHT (500 and 1000 µM) and 200 µl of β-carotene solution were added to wells in a 96-well flat-bottom microtitration plate. The sample mixture was diluted by transferring 30 µL to another plate containing air-sparged distilled water (210 µL). Dilutions were done in triplicate since the β-carotene bleaching reaction was subjected to noticiable variations. ADIBA (20 µL, 0.3 M) was added to each well to initiate the reaction. Absorbance readings were recorded every 10 min in a Multiskan Multisoft 349 plate reader (Labsystems) using a 450 nm filter at 0 min and at intervals 10 up to 90 min. Plates were kept in the dark at room temperature between readings. Antioxidant activity was calculated by four different methods previously described (*7, 14, 25, 26*). For the first method the log of the absorbance was plotted against time, as a kinetic curve, and the slope was expressed as the AOX value. The second method of calculation was based on first order kinetics and was conducted as described by Al-Saikan et al. (*14*):

$$\ln(a/b) \times 1/5 = \text{sample degradation rate} \qquad (1)$$

where ln is the natural log, a is the initial absorbance at 450 nm and at time 0; b is the absorbance at 450 nm and at 10, 20, and 30 min, and *5* represents the time in minutes. Antioxidant activity (AA) was also calculated as % inhibition relative to the control using the following equation (*14*):

$$AA = (R_{control} - R_{sample}) \times 100 / R_{control} \qquad (2)$$

where $R_{control}$ and R_{sample} are the degradation rates of β-carotene in reactant mix without and with sample extract, respectively. The AA values for different times were averaged to give one AA value for the sample.

The third method of expression is based on the oxidation ratio (ORR) and was calculated using the following equation (*25*):

$$ORR = R_{sample} / R_{control} \qquad (3)$$

where R_{sample} and $R_{control}$ are the same as in equation (2).

In the fourth method, the antioxidant activity coefficient (AAC) was calculated according to Mallet et al. (*25*):

$$AAC = (A_{sample\ 90} - A_{control\ 90}) \times 1000 / (A_{control\ 0} - A_{control\ 90}) \qquad (4)$$

where $A_{sample90}$ is the absorbance of the sample at $t = 90$ min, and $A_{control\ 0}$ the absorbance of the control at $t = 0$ min.

Statistical Analysis

Concentration of phenolic compounds, antioxidant activity and mutagenicity results represent the average and standard deviation (SD) from triplicate plates per dose for the two experiments performed in each test series. Statistical differences between control and treatment samples were analysed by Tukey's test.

Results and Discussion

Total Polyphenolic and Anthocyanin Compounds

Table I shows the concentration of condensed tannins and total anthocyanins. *P. coccineus* had the highest concentration of condensed tannins (925.55 ± 73 mg equivalents of (+)-catechin/g) while *P. vulgaris* (Negro Jamapa) had 621 ± 64 mg equivalents of (+)-catechin/g. The amount observed is in the range of our previous reports for *P. vulgaris* (5, 6, 27). For seed coat, we expected the difference to be higher, since the seed from *P. coccineus* is bigger than that of *P. vulgaris* (49.74 ± 4 and 197.72 ± 17 mg equivalents of (+)-catechin/g of seed coat, respectively). Differences between these results and our

Table I. Concentration of Phenolic Compounds in Common Beans *Phaseolus vulgaris* (Negro Jamapa) and *Phaseolus coccineus* (Ayocote Negro)

Cultivar	Whole Bean (mg/g)	Seed Coat (mg/g)	Phenolic Extract from Seed Coat (mg/g)
Condensed tannins[1]			
Phaseolus vulgaris	4.98 ± 0.35[a]	49.74 ± 2.4[a]	621 ± 64[a]
Phaseolus coccineus	17.6 ± 0.65[b]	197.72 ± 17[b]	925.55 ± 73[b]
Total anthocyanins[2]			
Phaseolus vulgaris	0.56 ± 0.02[a]	5.65 ± 0.25[a]	60.74 ± 4.25[a]
Phaseolus coccineus	0.17 ± 0.04[b]	1.9 ± 0.15[b]	8.64 ± 0.43[b]

The results represent the average of three independent experiments ± SD. Means with different superscript letters are significantly different (Tukey $\alpha = 0.05$). [1]Quantification of condensed tannins was done according to Desphande et al. (1985) and Desphande and Cheryan (1987), as described in Materials and Methods. mg/g = mg of (+)-catechin equivalent g/sample material. [2]Quantification of total anthocyanins was done according to Abdel and Hucl (1999), as described in Materials and Methods. mg/g = mg of cyanidin 3-glucoside equivalent/g sample material.

previous reports could be due to the fact that we are testing different species of beans and these cultivars were grown in different environments. It is also possible that the phenolic compounds are secondary metabolites that are influenced by the environmental conditions. The anthocyanins concentration in seed coat for *P. vulgaris* (criollo) was 5.65 ± 0.25 mg/g of seed coat and 1.9 ± 0.15 mg/g of seed coat for *P. coccineus,*, which is lower than the values reported for black bean (*P. vulgaris* L) (28) and *P. vulgaris* L Jamapa cultivar (27). This difference may indicate that the amount of anthocyanins is dependent on the variety of bean.

Protein Content and *in vitro* Protein Digestibility

Protein content of species of beans is shown in Table II. It is observed that the Negro Jamapa seed bean has a higher protein content ($p<0.05$) than Negro Ayocote. However, absolute differences between protein contents of both species are minimal. Protein levels were in the range reported by other investigators (29). On the other hand, protein digestibility is lower ($p<0.05$) in raw than in cooked beans. Protein digestibility levels are in agreement with the results reported by Guzman et al. (30). It is interesting to note that in spite of lower protein digestibility of raw seed of Negro Ayocote, cooked bean from both species showed similar ($p<0.05$) digestibility.

Table II. Protein, Water Absorption, Cooking Time and *in vitro* Digestibility in Common Beans *Phaseolus vulgaris* (Negro Jamapa) and *Phaseolus coccineus* (Ayocote Negro) Cultivars[1]

Bean Type	Protein (%) N x 6.25	In vitro Digestibility (%)		Water Absorption (%)	Cooking Time (min)
		Uncooked seed	Cooked seed		
Phaseolus vulgaris	17.94 ± 0.4^a	76.28 ± 1.0^a	96.58 ± 2.1^a	101.74 ± 5.2^a	90^a
Phaseolus coccineus	17.41 ± 0.02^b	69.24 ± 0.3^b	90.64 ± 1.2^a	119.65 ± 7.7^b	150^b

[1]Data shown are means of at three independent experiments $\pm$ SD. Means with different superscript letters are significantly different (Tukey $\alpha = 0.05$).

Water Absorption and Cooking Time

Neither species showed "hard to cook" or "hard shell" symptoms given that their seeds showed higher water absorption capacity and lower cooking time (Table II). Hard to cook and hard shell problems have been extensively reported as major causes of the long cooking time for common beans (*18*). It has also been reported that beans with higher water absorption capacity show lower cooking time. When beans absorb 10% more water, the cooking time decreases by approximately 10 min (*17*).

Antimutagenicity Assay

The dose-response curves of AFB1 mutagenicity in tester strain TA100 and TA98 are shown in Figure 1. Doses tested (0.125, 0.25, 0.5 and 1 μg/tube) were not toxic to the bacteria, and a dose of 0.5 μg/tube was chosen for subsequent antimutagenicity assays. The antimutagenicity effect of phenolic extract from *P.*

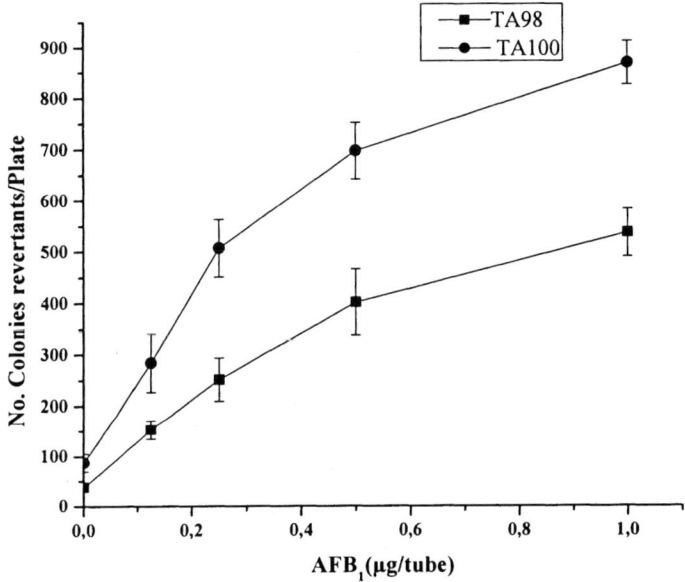

Figure 1. Dose-response curve of AFB$_1$ mutagenicity in TA98 and TA100. Each data point represents the mean of two independent experiments ± SD. The spontaneous mutation rate was 87 ± 18 and 38 ± 6 revertants/plate for TA100 and TA98, respectively.

Table III. Antimutagenic Effects of Phenolic Compounds Present in Methanolic Extracts from Common Beans *Phaseolus vulgaris* (Negro Jamapa) and *Phaseolus coccineus* (Ayocote Negro)[1]

Dry bean (mg)	Extract [μg eq. (+)-catechin/tube]	TA98		TA100	
		Revertants/ tube	Inhibi-tion (%)	Revertants/ tube	Inhibi-tion (%)
		P. vulgaris			
0	0	219 ± 28^a	0	426 ± 66^a	0
2	10	115 ± 6^b	47	221 ± 23^b	48
10	50	76 ± 9^c	65	$190 \pm 46^{b,c}$	55
20	100	$61 \pm 8^{c,d}$	72	$154 \pm 40^{b,c}$	64
40	200	$53 \pm 5^{d,e}$	75	$166 \pm 36^{b,c}$	61
80	400	44 ± 7^e	80	127 ± 36^c	70
		P. coccineus			
0.00	0	298 ± 36^a	0	402 ± 48^a	0
0.64	10	103 ± 19^b	65	173 ± 23^b	57
2.56	50	$75 \pm 20^{c,d}$	75	136 ± 22^c	66
5.13	100	$59 \pm 5^{c,d}$	80	$125 \pm 13^{c,d}$	69
10.25	200	$51 \pm 9^{d,e}$	83	$116 \pm 9^{c,d}$	71
20.5	400	41 ± 7^e	86	92 ± 4^d	77

[1]Results are the average of two independent experiments. Triplicate plates were tested per dose per experiment. Spontaneous mutation was 138 ± 8 revertants/tube for TA100 and 52 ± 11 revertants/tube for TA98. Different letters in the same column denote a statistically significant difference (Turkey, $\alpha = 0.05$). In all cases the mutagen was Aflatoxin B_1 (0.5 μg/tube).

vulgaris (criollo) and *P. coccineus* Ayocote negro, against AFB1 mutagenicity in TA98 and TA100 is shown in Table III. The phenolic extracts were neither toxic nor mutagenic to the bacteria at the concentrations tested because a consistently characteristic number of spontaneous revertants was observed for each tester strain (138 ± 8 revertants/tube for TA100 and 52 ± 11 revertants/tube for TA98, respectively). Although all the concentrations tested (10 to 400 μg equivalents (+)-catechin/tube) had an inhibitory effect on AFB1 mutagenicity in both tester strains, the response was dose-dependent for *P. vulgaris* at the lower concentrations tested (10 to 100 µg equivalents (+)-catechin/tube). However, the higher concentrations tested (200 and 400 µg equivalents (+)-catechin/tube) caused no significant increase in inhibition. Our results are in agreement with those obtained by other researchers using different cultivars of common bean (*6, 27*). *P. coccineus* (Ayocote Negro) had higher inhibition than *P. vulgaris* (Negro Jamapa) at the same concentration. For example, in TA98 with 10 µg equivalents (+)-catechin of phenolic extract of *P. coccineus*, the inhibition was 65%, while for common bean it was 47%. For a non-conventional species of bean such as *P. coccineus*, this is the first report to characterize its nutraceutical potential.

Antioxidant Activity

The β-carotene bleaching method was used to evaluate antioxidant activity (Table IV). BHT was used as standard at 500 and 1000 µM. *P. vulgaris* and *P. coccineus* were adjusted 500 and 1000 µM as µM catechin equivalents. The antioxidant activity was expressed in four different indices AOX (antioxidant activity), AA (% inhibition relative to control), ORR (ratio), and AAC (antioxidant activity coefficient). The inhibition of β-carotene bleaching by bean extracts showed lower antioxidant activity than the BHT. However, *P. vulgaris* (Negro Jamapa) extract had higher inhibition than *P. coccineus* (Ayocote Negro) at the concentrations tested. For example, for 1000 µM of *P. vulgaris* the AOX was 5.20; while for *P. coccineus* it was 6.56 (1.34 and 1.7 times higher than BH, respectively). Cardador-Martínez et al. (*7*) reported that phenolic extract from *P. vulgaris* Flor de Mayo cultivar had less antioxidant activity than BHT.

Conclusions

P. vulgaris (Negro Jamapa) and *P. coccineus* (Ayocote Negro) show antimutagenic and antioxidant activity. However, *P. coccineus* which is a non-conventional bean cultivar, had a higher concentration of condensed tannins, less *in vitro* digestibility, and higher antimutagenic activity than *P. vulgaris*. Due to their composition and biological activities, both cultivars could be suggested as good nutraceutical and functional foods. These results may indicate that a diet rich in beans and bean products may be useful in protecting against diseases and pathologies in which free radical production plays a key role. However, further research is necessary to fully characterize the health benefits that may be associated with consumption of these non-traditional bean varieties.

Table IV. Antioxidant Value (AOX), Antioxidant activity (AA), Oxidation Ratio (ORR), and Antioxidant Activity Coefficient (AAC) of Phenolic Compounds in Common Beans *Phaseolus vulgaris* (Negro Jamapa) and *Phaseolus coccineus* (Ayocote Negro)

Sample $(\mu M)^2$	Antioxidant activity, β-carotene method[1]			
	AOX	AA	ORR	AAC
	BHT^3			
1000	3.86^a	65.11^e	0.33^a	358.00^f
500	4.39^a	62.04^e	0.36^a	274.12^e
	Phaseolus vulgaris			
1000	5.20^b	58.35^d	0.44^b	187.43^d
500	5.76^b	51.04^b	0.51^c	136.51^c
	Phaseolus coccineus			
1000	6.56^c	54.18^c	0.50^c	98.19^b
500	6.92^d	45.25^a	0.60^d	51.07^a

[1]Means in a column followed by the same letter are not significantly different by Turkey's multiple range test ($\alpha = 0.05$), n = 3. $^2\mu M$ = (+)-catechin equivalent. [3]BHT = butylated hydroxytoluene standard.

References

1. Castellanos, J.Z.; Guzmán-Maldonado, S.H.; Jiménez, A.; Mejía C.; Muñoz-Ramos, J.J.; Acosta-Gallegos, J.A.; Hoyos, G.; López-Salinas, E.; González, E.D.; Salinas-Pérez, R.; González-Acuña, J.; Muñoz-Villalobos, J.A.; Fernández-Hernández, P.; Cásares, B. *Arch. Lainoam. Nutr.* **1997**, *27*, 163-167.
2. Beninger, C.W.; Hosfield, G.L. *J. Agric. Food Chem.* **2003**, *51*, 7879-7883.
3. Hangen, L.; Bennink, M.R. *Nutr. Cancer* **2002**, *44*: 60-65.
4. Rizkalla, S.W.; Bellisle, F.; Slama, G. *Brit. J. Nutr.* **2002**, *88*(Suppl 3): S255-S262.
5. González de Mejía, E.; Castaño-Tostado, E.; Loarca-Piña, G. *Mut. Res..* **1999**, *441*, 1-9.
6. Cardador-Martínez, A.; Cataño-Tostado, E; Loarca-Piña, G. *Food Addit. Contam.* **2002**, *19*, 62-69.
7. Cardador-Martínez, A.; Loarca-Piña, G.; Oomah, B.D. *J. Agric. Food Chem.* **2002**, *50*: 6975-6980.
8. Rockfound 2001. http://rockfound.org.mx/coccineusbiesp.html.

9. Calderón, E.; Velásquez, L.; Bressani, R. *Arch. Latinoam. Nutr.* **1992**, *42*: 64-71.

10. Pérez-Herrera, P.; Esquivel-Esquivel, G.; Rosales-Serna, R.; Acosta-Gallegos, J. *Arch. Latinoam. Nutr.* **2002**, *52*:172-180.

11. Nowosielski, J.; Podyma, W.; Nowosielska, D. *Cell. Mol. Biol. Lett.* **2002**, *7*(2B), 753-762.

12. Deshpande, S.S.; Cheryan, M. *J. Food Sci.* **1985**, *50*, 905-910.

13. Deshpande, S.S.; Cheryan, M. *J. Food Sci.* **1987**, *52*, 332-334.

14. Al-Saikan, M.S.; Howard L. R.; Miller, J.C., Jr. *J. Food Sci.* **1995**, *60*, 341-343.

15. Guzmán-Maldonado, S.H.; Castellanos, J.; González de Mejía E. *Food Chem.* **1996**, *55*: 333-335.

16. *AOAC Official Methods;* AOAC International: Gaithersburg, MD, 1997; Vol. II, 15th edition, Methods 955.04 Nitrogen total, and 982.30 Protein efficiency.

17. Castellanos, J.Z.; Guzmán-Maldonado, S.H.; Acosta-Gallegos, J.A.; Kelly, J.D. *J. Sci. Food Agric.* **1995**: *69, 437*.

18. Reyes-Moreno, C.; Paredes-López, O. *CRC Crit. Rev. Food Sci. Nutr.* **1993**, *33*: 227-286.

19. Kado, N.Y.; Langley, D.; Eisenstadt, E. *Mut. Res.* **1983**, *121*, 25-32.

20. Kado, N.Y.; Guirguis, G.N.; Flessel, C.P.; Chan, R.C.; Chang, K.; Wesolowski, J.J. *Environ. Mutagen.* **1986**, *8*, 53-66.

21. Ames, B.N.; McCann, J.; Yamasaki E. *Mut. Res.* **1975**, *31*, 347-364.

22. Fukumoto, L.R.; Mazza, G. *J. Agric. Food Chem.* **2000**, *48*, 3597-3604.

23. Marco, G.J. *J. Am. Oil Chem. Soc.* **1968**, *45*, 594-598.

24. Velioglu, Y.S.; Mazza, G.; Gao, L.; Oomah, B.D. *J. Agric. Food Chem.* **1997**, *45*, 304-309.

25. Mallet, J.F.; Cerrati, C.; Ucciani, E.; Gamisana, J.; Gruber, M. *Food Chem.* **1994**, *49*, 61-65.

26. Marinova, E.M.; Yanishlieva, N.; Kostova, I.N. *Food Chem.* **1994**, *51*, 125-132.

27. Aparicio-Fernández, X.; Manzo-Bonilla, L.; Loarca-Pina, G. *J. Food Sci.* **2005**, *70*, S73-S78.

28. Takeoka, G.R.; Dao, L.T.; Full, G.H.; Wong, R.Y.; Harden, R.A.; Edwards, R.H.; Berrios, J.D. *J. Agric. Food Chem.* **1997**, *45*, 3395-3400.

29. González de Mejía E.; Martínez-Resendiz, V.; Castaño-Tostado, E.; Loarca-Piña G. *J. Sci. Food Agric.* **2003**, *83*, 1022-1030.

30. Guzmán-Maldonado, S. H., Acosta-Gallegos, J.; Paredes-López, O. *J. Sci. Food Agric.* **2000**, *80*, 1874-1881.

Chapter 9

Amaranth: An Ancient Crop for Modern Technology

Ana Paulina Barba de la Rosa[1], Cecilia Silva-Sánchez[1],
and Elvira González de Mejia[2]

[1]IPICyT, Camino a la Presa, San José No. 2055, Col. Lomas 4ta,
C. P. 78216, San Luis Potosí, Mexico
[2]Department of Food Science and Human Nutrition, University of Illinois
at Urbana-Champaign, 1201 West Gregory Drive, 228 ERML, MC–051,
Urbana, IL 61801

Amaranth may play a very important role in the fight against
hunger and contribute to the wellness of people. Amaranth
contains, besides nutritional benefits, compounds with
potential nutraceutical properties against hypertension and
cancer. New technologies and agricultural strategies must be
developed to increase the availability of this important novel
food alternative.

Amaranth is a pseudocereal which provides both grains and tasty leaves of high
nutritional value. Due to its properties, it can be considered as a functional food.
The seed has a high protein content and its amino acid composition is closer to
the optimum balance required in the human diet. The leaves also contain high
levels of proteins (27.8 to 48.6%), unsaturated oil (45% of linoleic acid of total
fat), fiber (11.1 to 23.2%) and minerals such as potassium, iron, magnesium and
calcium (1). Aside from these nutritional components, amaranth also contains
other compounds that play important biological roles, such as protease
inhibitors, antimicrobial peptides, lectins, and antioxidative compounds

(*2-4*). Aqueous extracts of *Amaranthus gangeticus* leaves have been reported to possess anticancer activity in liver, breast, and colon cancer cell lines (*5*). Leaves of *Amaranthus tricolor* were also found to have anti-tumor and anti-cell proliferation activities as well as presence of cyclooxygenase enzyme inhibitory compounds (*6*). Recent reports have shown that amaranth seed storage proteins contain anticarcinogenic and antihypertensive peptides (*7*).

History

Amaranth, though relatively unknown today, was cultivated as a highly nutritious plant that was a dietary staple in pre-Columbian Mesoamerica as early as 7000 years ago. Amaranth grain was an important source of protein and vitamins to the Aztecs who called it "huatli." Along with beans and corn, amaranth was a fundamental part of the indigenous Mexican diet (*8, 9*). Like the Aztecs, the Mayas and the Incas also cultivated and revered amaranth (Fig. 1A). In accordance to the Aztec calendar, popped amaranth grain (Fig. 1B) was ground and mixed with honey, and this paste would then be formed into the shapes of serpents, birds, or mountains. The sacred amaranth figurines were paraded through the streets to the grand temples of Tenochtitlan where they were broken into pieces and distributed to the people. Due to its ritual use the conquering Spaniards banned amaranth cultivation and consumption in the 16th century. As a result, domesticated amaranth nearly disappeared from Mexico (*9*). However, a handful of communities in Mexico continued to grow amaranth and today refer to the sacred plant of their ancestors as "alegria" (happiness). However, the positive attributes of the crop led to its adoption in other areas around the world. By the 1700s, amaranth had spread throughout Europe for use as an herb and ornament. In the late 1800s, amaranth was reportedly being grown in mountain valleys of Nepal and parts of East Africa. During the 20th century, because of its impressive nutritional qualities and desirable agricultural properties, which include drought resistance and amenability to cultivation in temporal fields, amaranth has captured the interest of researchers from many countries including the US, Australia, China, India, and several African nations (*8*). Due to all of its nutritional benefits, Amaranth seed (often referred to as grain) contains a nearly perfect balance of those essential amino acids that the human body needs to make protein. The essential amino acid lysine, which is scarce in all other cereal grains, is abundant in amaranth (*10*).

The leaves are high in calcium, iron, and folic acid. Amaranth's nutritional properties make it an ideal supplement of the traditional Mexican diet especially in the marginalized poor communities. The consumption of amaranth could play a very important role in the fight against hunger and malnutrition (*11*).

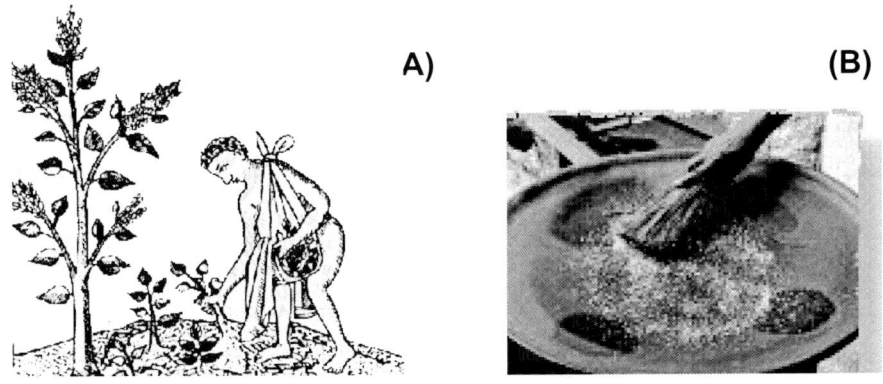

Figure 1. (A) Amaranth recollector. (B) Traditional amaranth popping.

Agronomical Features

Amaranth belongs to the *Amaranthacea* family. The genus *Amaranthus* has 60 species, but only three of them are fit for human consumption: *A. hypochondriacus, A. caudatus,* and *A. cruentus. A. hypochondriacus* and *A. cruentus* are cultivated in Mexico (Fig. 2), and *A. caudatus* is grown in South America (*8*). Amaranth has the ability to prosper in regions with little or erratic precipitation, which makes it an excellent alternative for temperate zones, as it needs less than half the water that most grains require. The conditions for he optimal growth may differ according to the species cultivated. In general, Amaranth grows better when the temperatures are up to 21°C and the temperatures of germination are between 16 to 35°C. The culture is developed in atmospheres with a very ample variation that goes from 300 to 2000 mm of annual precipitation, in altitudes from sea level to 3000 m (Table I). Amaranth seeds are sufficiently strong and adequate to prosper in deficient soils where conventional grains are not easily cultivated (*12*). Amaranth has been described as plant-enhanced remediation of soil, specially *Amaranthus tricolor* enhanced the dissipation of the polycyclic aromatic hydrocarbons, and has been used as phytoremediation of soils contaminated with heavy metals (*13, 14*). Amaranth has high oxalic content (up to 10%) and 2-3% glucosides; it has been used as a biological preservative in silage, showing no signs of growing mold and rot (*15*). Amaranth is a very efficient plant, one of the few where the first product of photosynthesis is a four carbon compound. The combination of anatomical features in amaranth and C4 metabolism results in increased efficiency of CO_2 use under a wide range of both temperature and moisture stress environments, and contribute to the plants' geographic adaptability to diverse environmental

106

Figure 2. Two different Amaranthus hypochondriacus species. Crops containing white-cream seeds and purple seeds.

conditions (*8*). Table I presents the characteristics of *Amaranth hypochondriacus* grown at different locations (*16-18*).

Nutritional Value

Amaranth may be one of the most nutritious plants in the world. The nutritonal composition of both grain and leaves has been extensively studied, especially with regard to its high content of protein, calcium, folic acid, and vitamin C (*10, 19-21*). Popped amaranth seeds provide a good source of protein, which can satisfy a large portion of the recommended protein requirements for children and provide approximately 70% of necessary calories (Table II). In addition, a combination of rice and amaranth in a 1:1 ratio have been designated as an excellent way to achieve the protein allowance recommended by the World Health Organization (*22*). It is well known from the nutritional point of view that animal food proteins are superior compared with vegetable protein sources. However amaranth is one of the vegetable protein sources that is close to animal protein quality.

There is agreement in the literature that the popped seed has lower protein quality than the raw material, although the digestibility *in vivo* and *in vitro* is slightly higher when compared with other cereals or even with animal protein sources (*e.g.* casein). Protein quality parameters like the protein efficiency ratio (PER) of raw or popped seeds is superior to wheat, and the raw material is close to the casein reference point (*23, 24*).

Nutraceutical Properties of Amaranth Flours

In addition to the nutrients that are involved in normal metabolic activity, plant foods contain components that may provide other health benefits. One health niche for amaranth protein flours is as a gluten- or prolamin-free protein; celiac disease is characterized by sensitivity to the prolamin fraction of cereals (*25*). It has been reported that amaranth inhibits antigen-specific IgE production through augmentation of the interferon-gamma response *in vivo* and *in vitro*, meaning that the immune system is stimulated by a component present in amaranth (*26*). Another issue about the nutraceutical value of amaranth-based diets is its hypocholesterolemic effect. It has been observed that amaranth positively affects the plasma lipid profile in rats fed cholesterol-containing diets, suggesting that this positive influence may be due to the presence of bioactive peptides and antioxidants (*27-30*). In general, several bioactive peptides have

Table I. Characteristics of *Amaranth hypochondriacus* Grown in Different Locations in Mexico

	Oaxaca	Guanajuato	San Luis Potosí
Race or variety	Azteca	Nutrisol	Nutrisol
Color of plant and inflorescence	mix of colors (green, salmon, purplish-red)	white-cream	white-cream
Height of plant (m)	1.8 to 2.5	1.2 to 1.7	1.4 – 1.6
Type of inflorescence	erect and compact	erect and compact	erect and compact
Size of inflorescence (cm)	60 to 110	215	180
Cultivation cycle (d)	180	180	180
Growing season	April to July	May to August	May to September
Grain yield (kg/Ha)	1000	1809	1121
Reference	(*16*)	(*17*)	(*18*)

Table II. Nutritional Composition of Amaranth Seeds Compared with Several of the Most Important Food Crops (Based on 100 g Samples)

	Amaranth	Rice	Wheat	Corn	Oats
Protein (%)	19	5.6	12.8	9.4	15.8
Fiber (%)	5.6	0.3	2.3	3	3
Fat (%)	6	0.6	1.7	4.7	6.9
Carbohydrate (%)	60	79.4	71	74	66
Calcium (mg)	250	9	29.4	7	54
Iron (mg)	15	4.4	4	2.7	5
Calories (kcal)	414	360	334	365	389

References: 1, 12.

109

been found. Using bioinformatics tools (*31*), several bioactive peptides in amaranth have been detected *in silico* (Fig. 3). Thirty-six amaranth proteins were found in the database (www.ncbi.nlm.nih.gov). These sequences were tested for more than 1573 active peptides reported (www.uwm.edu.pl/biochemia) with 39 different activities. For amaranth, seed proteins were found to have antithrombotic, inmunomodulating, opioid, regulating, antioxidant, activating ubiquitin-mediated proteolysis (AUMP), protease inhibitor and antihypertensive activities. The peptides with antihypertensive activity were the most frequent in amaranth proteins (Fig. 3). These were followed by protease inhibitor activity, opioid and finally by activating ubiquitin-mediated proteolysis (*7*).

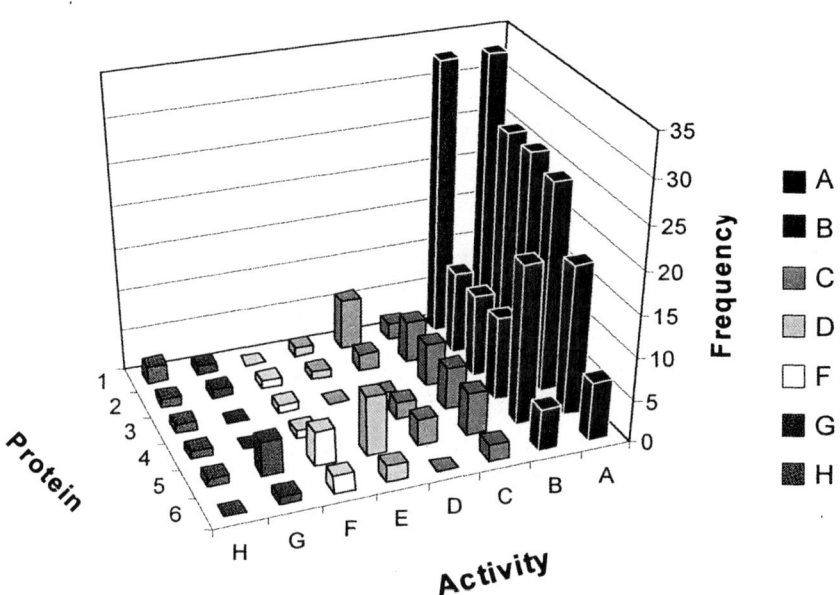

Figure 3. Main active peptides from amaranth proteins. Protein sequences:
1=11S globulin, 2=Chain α-agglutinin complex, 3=amaranth seed storage
protein, 4=35 kDa albumin, 5=chain A– alpha amylase, 6=trypsin inhibitor.
Peptide activities: A=antihypertensive, B=protease inhibitor, C=opioid,
D=activating ubiquitin mediated proteolysis, E=regulating ion flow and
stomach membrane activity, F=immunomodulating, G=antithrombotic,
H=antioxidative.

Amaranthus spinosus Linn. (thorny amaranth) is extensively used in Chinese traditional medicine to treat diabetes and it has also shown immunostimulating activity via direct stimulation of B lymphocyte activation *in vitro* (*32*).

Amaranth extract may have therapeutic effects and gastroprotection against 100% ethanol, acting on gastric mucosa stimulating afferent nerves and increasing gastric microcirculation (*36*).

Another nutraceutical component of amaranth is the recently found lunasin-like peptide. Lunasin is a unique 43 amino acid peptide found in soybean (*33, 34*). It has been demonstrated that transfection of mammalian cells with lunasin gene arrest mitosis, leading to cell death (*35, 36*). The lunasin peptide was originally cloned, purified, and chemically synthesized from soybean; since then only barley has been identified as a natural alternative source of this peptide (*37, 38*). Significant amounts of lunasin have been found in albumin, globulin 7S, globulin 11S, and amaranth fractions. The glutelin fraction showed the highest lunasin content, amounts ranging from 2.72 to 3.11 µg of lunasin/g total protein.

It is concluded that the amaranth proteins could be an alternative source of bioactive peptides, and particularly lunasin can be beneficial to health (*7*).

Uses of Amaranth

The amaranth market remains relatively small and undeveloped, in part because there is a general lack of familiarity with it in the public and private sectors. To achieve a higher level of market penetration, amaranth will have to become more publicized, prices will have to fall (although a premium could still be commanded), and availability will have to be increased. Distance to buyers is a problem for many current amaranth growers. Special markets such as the starches or other seed components could lead to increase marketing opportunities. Altough amaranth has a variety of potential uses, its only commercial market in the U.S. and Mexico has been as added ingredient of organic food products. A few small companies have been the primary buyers and processors of amaranth, including Arrowhead Mills (Texas), Health Valley (California), American Resources (Minnesota), San Miguel de Proyectos (México), and Grupo Agroindustrial de Alimentos Mexicanos (Mexico). Due to their great adaptability to adverse conditions, cultivation of amaranth is becoming as an alternative crop in Northern Europe (*9, 39*). Popping is a simple processing of amaranth seeds for food uses, an example of this processing is the candy called "alegria" in Mexico. A fluidized bed system, for popping of amaranth, has been developed by Konishi et al (*40*). It is believed that this method protects the nutritional value of the seed.

Market demand for amaranth has fluctuated over the last decade, but there has been steady use of the crop for breakfast cereals, and more recently, in mass produced multi-grain bread products. Several products have been developed using amaranth flour as main component. Such products are extruded snacks (*41, 42*), amaranth pasta (*43*), and bread with up to 20% of amaranth flour substituted *44*) or added with amaranth albumins (*45*). Most typically, amaranth products are sold in the health food sections of grocery stores, in specialty food stores, or through direct marketing. Amaranth to date has only appeared in mainstream products when used as a minor component of multigrain foods (*46*).

Recently a biofilm based on amaranth flour was developed. The aim was to test the filmogenic capacity of amaranth to make edible films. The films had better barrier properties and good flexibility, but were slightly yellowish and the mechanical resistance was relatively low as compared to those made from different sources such as wheat and soybean (*47*).

The amaranth leaves have been studied as a valuable material for production of food additives and enriched tea products, due to their high content of vitamins, flavonoids, pectins, amaranthin, and other trace elements (*48-50*). The importance of leafy vegetables (including leaves from cultivars like pumpkins, melons, and cowpeas; as well as leaves from wild and weedy species like amaranth) consumed by indigenous populations in sub-Saharan AfricaIt has also been reported, suggesting that they could be important sources of micro nutrients including vitamin A and C and iron among others (*11*). A study conducted by Haskell et al. (*51*) in Nepal on pregnant woman, tested the treatment of night blindness with supplementation of vitamin A from food or synthetic sources that include retinyl palmitate, vitamin A-fortified rice, goat liver, amaranth leaves, or carrots. They concluded that improvement in dark adaptation did not differ significantly among women who received vitamin A as liver, amaranth leaves, carrots, or retinyl palmitate. Thus, in populations where the food is limited, amaranth leaves can be an alternative to improve health.

Amaranth Starch Granules

Starch is used nowadays as a renewable raw material, as a source of energy after conversion to ethanol, and for many different industrial applications. Starch is a versatile and useful polymer, not only because of its low price. In recent years, the understanding of starch structure and our knowledge of the enzymes that are involved in starch biosynthesis has increased greatly; many of the genes that encode these enzymes have been cloned. Most crops are now amenable to transformation by *Agrobacterium tumefaciens*, and so it has become possible, using genetic modification techniques, to alter the expression of individual starch biosynthetic genes and to study the effect of such changes on starch structure.

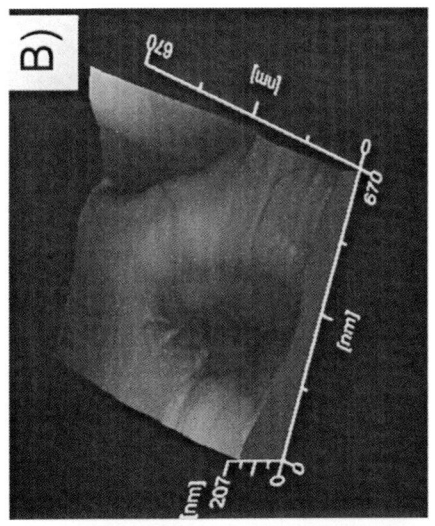

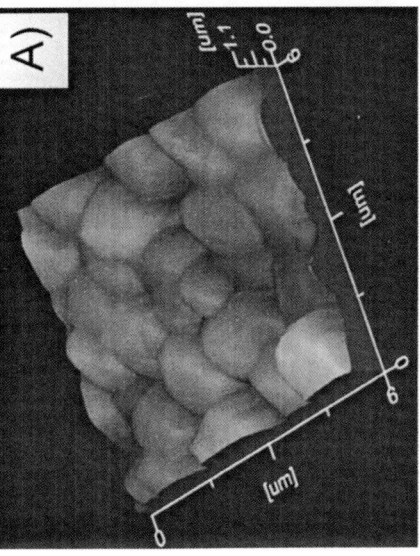

Figure 4. Atomic Force Microscopy (AFM) of amaranth starch granules. A) Starch granules size around 2 μm. B) Superficial holes on granules of approximately 300 nm.

There is now great potential for creating designer starches that have novel functionalities (*52*).

Among the starches, the granules from amaranth starch (*A. hypochondriacus*, amylopectin type) are special because of their extremely small size of 1-3 μm and high uniformity (Fig. 4). However, large spherical particles of 30-80 μm in diameter have been observed from spray-dried amaranth starch by environmental scanning electron microscopy (ESEM), which exhibited a characteristic fine structure. Recently, amaranth starch has raised scientific interest, but its characterization is limited so far to common properties such as solubility in cold water, swelling ratio, freeze-thaw stability, water retention capacity, and aqueous paste viscosity (*53*).

Feruloylated oligosaccharides show that ferulic acid is predominantly bound to pectic arabinans and galactans in amaranth (*Amaranthus caudatus* L.) insoluble fiber (*54*).

Conclusions

The current status of amaranth is as a crop which has great potential and a variety of possible uses. However, as with most alternative crops, cultivar improvements are needed, production and utilization research challenges remain, and major barriers exist in market development.

Perhaps the end-use product will determine the key issues in the future of amaranth. If the primary amaranth product use will continue to be a niche for the organic health food market, then we do not foresee a major expansion of production acreage. However, if amaranth can be incorporated into a major flour milling blend for large volume uses, then significant expansion could follow. Its agronomic and nutraceutical high value will be taken into consideration for its future development.

References

1. Segura-Nieto, M.; Barba de la Rosa, A.P.; Paredes-López, O. In *Amaranth Biology, Chemistry and Technology;* Paredes López, O., Ed.; CRC Press: Boca Raton, FL, 1994; pp 76-106.
2. Broekaert, W.F.; Mariën ,W.; Terras, F.R.G; De Bolle, M.F.C; Proost, P.; Van Damme, J.; Dillen, L.; Claeys, M.; Ress, S.B.; Vanderleyden, J.; Cammue, B.P.A. *Biochem.* **1992**, *31,* 4308-4314.
3. Duarte-Correa, A.; Jokl, L.; Carlsson, R.. *Arch. Latinoam. Nutr.* **1986**, *36,* 319-326.

4. Valdés-Rodríguez, S.; Segura-Nieto, M.; Chagolla-López, A.; Verver y Vargas-Cortinas, A.; Martínez-Gallardo, N.; Blanco-Labra, A. *Plant Physiol.* **1993**, *103,* 1407-1412.
5. Sani, H.A.; Rahmat, A.; Ismail, M.; Rosli, R.; Endrini, S. *Asia Pac. J. Clin. Nutr.* **2004,** *13,* 396-400.
6. Jayaprakasam, B.; Zhang, Y.; Nair, M.G. *J. Agric. Food Chem.* **2004,** *52,* 6939-6943.
7. Silva-Sánchez, C.; Barba de la Rosa, A.P.; De Mejia Gonzalez, E.; 229th ACS National Meeting, 2005, Abstract AGFD96.
8. Stallknecht, G.M.; Schulz-Schaeffer, J.R. In *New crops;* Janick, J., Simons, J.E., Eds.; Wiley, New York, NY, 1993; pp 211-218.
9. Myers, R.L. In *Progress in New Crops;* Janick, J., Ed.; ASHS Press, Alexandria, VA, 1996; pp. 207-220.
10. Barba de la Rosa, A.P.; Gueguen, J.; Paredes-López, O.; Viroben, G. *J. Agric. Food. Chem.* **1994,** *40,* 931-936.
11. Jansen van Rensburg, W.S.; de Ronde, J.A.; Venter, S.L.; Netshiluvhi, T.R.; van den Heever, E.; Vorster, H.J. *S. Afr. J. Bot.,* **2004,** *70,* 52-59.
12. Paredes-López, O.; Barba de la Rosa, A.P.; Hernández, D.; Carabez, A. *Características Alimentarias y Aprovechamiento Agroindustrial.* Secretaría General de la Organización de los Estados Americanos. Programa Regional de Desarrollo Científico y Tecnológico. Washington, DC, 1991; pp 1-20.
13. Ling, W.; Gao, Y. *Environ. Geol.* **2004,** *46,* 553-560.
14. Antonkiewicz, J.; Jasiewicz, C. *Acta Sc. Pol. Form. Circum.* **2002,** *1,* 119-130.
15. Allaberdin, I.L. *Kormoproizvodstvo* **2004,** *9,* 31-32.
16. Centro de Desarrollo Comunitario Centéotl. Oaxaca, México, 2001, http://www.prodigyweb.net.mx/centeotlac/.
17. González-Castañeda, J.; Borodanenko, A.; Cabrera-Sixto, M.; Quiróz-Ramírez, C. *Amaranto su Cultivo y su Procesamiento.* Coordinación de Comunicación Social. Legislatura del Congreso del Estado de Guanajuato, Mexico, 2001; pp 1-48.
18. Olvera-Martínez, J.L.; Olivo-Preciado, J.; Barba, A.P.; Silva-Sánchez, C. XIV Congreso Nacional de Investigación y Desarrollo Tecnológico Agropecuario. Durango, Dgo. 2003, 1-14.
19. Becker, R.; Wheeler, E.L.; Lorenz, K.; Stafford, A.E.; Grosjean, U.K.; Betschart, A.A.; Saunder, R.M. *J. Food Sci.* **1981,** *46,* 1175-1180.
20. Teutonico, R.A.; Knorr, D. *Food Technol.* **1985,** *39,* 49-60.
21. Bressani, R. Grain Amaranth: Its chemical composition and nutritive value. In *Proc. Fourth Amaranth Symp.;* Minn. Ext. Serv., St. Paul, MN, 1990.
22. FAO/WHU/UNO. *Cereal Food World* **1986,** *3,* 694-695.
23. Gamel, T.H.; Linssen, J.P.; Alink, G.M.; Mosallem, A.S.; Shekib, L.A. *J. Sci. Food Agric.* **2004,** *84,* 1153-1158.

24. Gamel, T.H.; Linssen, J.P.; Mesallem, A.S.; Damir, A.A.; Shekib, L.A. *J. Sci. Food Agric.* **2005**, *85,* 319-327.
25. Robins, G.; Howdle, P.D. *Curr. Opin. Gastroenterol.* **2005,** *21,* 152-161.
26. Hibi, M.; Hachimura, S.; Hashizume, S.; Obata, T.; Kaminogawa, S. *Cytotechnol.* **2003,** *43,* 33-40.
27. Czerwinski, J.; Bartnikowska, E.; Leontowicz, H.; Lange, E.; Leontowicz, M.; Katrich, E.; Trakhtenberg, S.; Gorinstein, S. *J. Nutri. Biochem.* **2004,** *15,* 622-629.
28. Marcone, M.F.; Kakuda, Y.; Yada, R.Y. *Plant Foods Hum. Nutr.* **2003,** *58,* 207-211.
29. Shin, D.H.; Heo, H.J.; Lee, Y.J.; Kim, H.K. *Brit. J. Biomed. Sci.* **2004,** *61,* 11-14.
30. Berger, A.; Gremaud, G.; Baumgartner, M.; Rein, D.; Monnard, I.; Kratky, E.; Geiger, W.; Burri, J.; Dionisi, F.; Allan, M.; Lambelet, P. *Int. J. Vitam. Nutr. Res.* **2003,** *73,* 39-47.
31. Dziuba, J.; Iwaniak, A.; Minkiewicz, P. *Polimery* **2003,** *48,* 50-53.
32. Lin, B.F.; Chiang, B.L.; Lin, J.Y. *Int. J. Immunopharmaco.* **2005,** *5,* 711-722.
33. Odani, S.; Koide, T.; Ono, T. *J. Biol. Chem.* **1987,** *262,*10502-10505.
34. Jeong, H.J.; Park, J.H.; Lam, Y.; De Lumen, B.O. *J. Agric. Food Chem.* **2003,** *51,* 7901-7906.
35. Galvez, A.F.; De Lumen, B.O. *Nat. Biotech.* **1999,** *17,* 495-500.
36. Galvez, A.F.; Chen, N.; Macasieb, J.; De Lumen, B.O. *Cancer Res.* **2001,** *61,* 7473-7478.
37. Jeong, H.J.; Lam, Y.; De Lumen, B.O. *J. Agric. Food Chem.* **2002,** *50,* 5903-5908.
38. Gonzalez de Mejia, E.; Vasconez, M.; De Lumen, B.O.; Nelson, R. *J Agric. Food Chem.* **2004,** *52,* 5882-5887.
39. Jacobsen, S.E.; Itenov, K.; Mujica, A. *Agr. Trop. Maracay* **2002,** *52,* 109-119.
40. Konishi, Y.; Iyota, H.; Yoshida, K.; Inoue, T.; Nishimura, N.; Nomura, T. *Biosci. Biotechnol. Biochem.* **2004,** *68,* 2186-2189.
41. Chavez-Jauregui, R.N.; Cardoso-Santiago, R.A.; Pinto deSilva, M.E.M.; Areas, J.A.G. *Int. J. Food Sci. Technol.* **2003,** *38,* 795-798.
42. Majewski, Z.; Replinska, A. *Ann. Warsaw Agric. Eng.* **2002,** *42,* 69-73.
43. Kovacs, E.T.; Berghofer, E.; Schnonlechner, R. In *Flour-Bread 01: Proceedings of International Congress;* Ugarcic-Hardi, Z. Ed.; 3rd Croatian Congress of Cereal Technologists, Opatija, Croatia, 2002; pp 70-78.
44. Park, S.H.; Morita, N. *Food Sci. Technol. Res.* **2004,** *10,* 127-131.
45. Silva-Sanchez, C.; Gonzalez-Castañeda, J.; De Leon-Rodriguez, A.; Barba de la Rosa, A.P. *Plant Foods Human Nutr.* **2004,** *59,* 169-174.

46. Schnetzler, K.A.; Breene, W.M. In *Amaranth Biology, Chemistry and Technology*. Paredes López, O., Ed.; CRC Press, Boca Raton, FL, 1994; pp 155-184.
47. Tapia-Blácido, D.; Sobral, J.P.; Menegalli, F.C. *J. Food Eng.* **2005**, *67*, 215-223.
48. Kononkov, P.F.; Gins, M.S.; Rakhimov, V.M.; Gins, V.K.; Logvinchuk, T.M. *Kartofel. Ovoshchi.* **2004**, *1*, 29-30.
49. Anilakumar, K.R.; Khanum, F.; Sudarshanakrishna, K.R.; Santhanam, K. *Indian J. Exp. Biol.* **2004**, *42*, 595-600.
50. Zayachkivska, O.S.; Konturek, S.J.; Drozdowicz, D.; Konturek, P.C.; Brzozowski, T.; Ghegotsky, M. R. *J. Physiol. Pharmaco.* **2005**, *56*, 219-231.
51. Haskell, M.J.; Pooja-Pandey, G.J.M.; Peerson, J.M.; Shrestha, R.K.; Brown, K.H. *Am. J. Clin. Nutr.* **2005**, *81*, 461-471.
52. Jobling, S. *Curr. Opin. Plant Biol.* **2004**, *7*, 210-218.
53. Wilhelm, E.; Aberle, T.; Burchard, W.; Landers, R. *Biomacromolecules* **2002**, *3*, 17-26.
54. Bunzen, M.; Ralph, J.; Steinhart, H. *Mol. Nutr. Food Res.* **2005**, *49*, 551-559.

Chapter 10

A Comparative Analysis of Lime Species and Flavor Relevance in Hispanic Food

Suzanne C. Johnson

Premium Ingredients International, 285 East Fullerton Avenue,
Carol Stream, IL 60188

The flavor of lime is prevalent throughout Hispanic food and food products but until recently has found somewhat limited application within the U.S. market. The increasing U.S. Hispanic population has driven a rapid increase in the number and variety of lime flavored food products introduced to the U.S. market. This includes products flavored with lime alone or in combination with other flavors. While some products are targeted to the Hispanic consumer, many are more mainstream. The typical profile of lime flavor found in U.S. products is quite different from the typical Hispanic limon flavor. It is a critical distinction that must be understood when formulating products to appeal to a particular market segment.

There are many reasons why we eat and why we choose to eat the way we do. Our basic nutritional needs could be covered by a diet far different from that consumed by most of us. The preference for certain foods and flavors is a relatively complex matter affected by our emotions and memories as well as our basic senses. There is also the undeniable cultural influence that greatly influences our choices. It can be said that familiarity leads to preference. When a food product that has been familiar and consumed frequently is removed from our diet, we have a tendency to crave that flavor. When this happens, only the exact product will satisfy our craving, and in fact a substitute that is similar is often rejected as undesirable. While an individual may develop a fondness for the new version, inevitably it will never be as good as the original. Similarly, when someone has a great fondness for a particular food product (mango, for example), it is not uncommon that they find an artificially flavored mango product to be offensive. This concept translates directly to the typical Hispanic vs. typical American expectation of lime flavor.

Lime Flavor

Lime has always been a prevalent flavor within the Hispanic food market, from beverages to candy, chips to snacks and cakes, and of course alcoholic beverages as evidenced by the popular corona with a wedge of lime and the ever present margarita. Within the last few years the U.S. market has seen a huge growth in the number of products flavored with lime, either stand-alone or in combination with other flavors. Beverages, potato and tortilla chips, yogurts, and candy are just the beginning. Some have targeted the growing Hispanic population, while others are mainstream products with a twist. As the U.S. becomes more exposed to Hispanic products and foods there is a growing demand not only for authentic Hispanic products but also for fusion type products to appeal to the mass market. One critical prerequisite for successful product development in this arena is an understanding of flavor profile and its appeal to the target market.

Lime is a prevalent flavor throughout Hispanic culture and until recently has found relatively little application in the U.S. market. It is also a great example of the difference in flavor expectations. The typical Hispanic lime profile is acidic and zesty, and is often served in combination with heat or spice in both food and candy applications. Even the Hispanic lime-flavored beverages are much different in flavor from the sweeter, candy-like profile typically found in the U.S. This is due in part to the prevalent variety of lime (Table I).

In a product such as lime, both acidity (tartness) and sweetness, as well as salt if applied, are as critical to the flavor perception as the aroma chemicals which inherently make a lime taste like lime. A lime flavor that tastes naturally tart and juicy in combination with acid and sugar can become sweet and candy-

Table I. Comparison of Common Lime Varieties

Citrus aurantifolia	*Citrus latifolia*
Key Lime	Persian Lime
Mexican Lime	Tahitian Lime
West Indian Lime	Bearss Lime
	Limon persa (Mexico)
	Limon sutil (South America)
Grown in Mexico, Peru, Haiti	Grown in Florida, Brazil
Smaller fruit, contains seeds	Larger seedless fruit
Yellow when ripe	Green when ripe, except Bearss turns yellow under appropriate ripening conditions
Higher acid content, stronger flavor	Lower acid content, weaker flavor
Sensitive to cold	Enhanced resistance to cold

like when the acid is decreased. With lime being especially susceptible to this phenomenon, the composition of aroma chemicals can be contradicted by incorrect acid type and level in the final product.

The Lime Fruit

The fruit itself is composed of three distinct parts, two of which are important to flavor. The outer skin, otherwise known as the epicarp or flavedo, contains not only carotenoids that give the fruit its distinct color but also the oil sacs containing characteristic essential oils. These essential oils are typically extracted and used in flavor manufacture. The white spongy layer immediately under the outer skin is referred to as the mesocarp or albedo and does not play an important role in flavor except perhaps for some occasional bitterness. The inside of the fruit, the endocarp, is composed of membrane-separated segments. Inside these segments are vesicles containing the juice.

Citrus latifolia

Citrus latifolia, also known as the Tahitian Lime or Persian Lime, is the primary lime grown in the U.S. and it sold simply as lime. It was developed as a hybrid for its resistance to disease, pests, and climate. It is similar in size and

120

shape to a lemon. Although it is typically sold when green, it does turn yellow when fully ripe. It has a thick rind, is quite hardy, and has very few if any seeds.

Citrus aurantifolia

Citrus aurantifolia is also known as Key Lime, West Indian Lime, or Mexican Lime. Most of the Key Limes found in the U.S. are currently imported from Mexico where it is grown year round. Within the U.S., they are grown in Florida, Texas, and California where they are seasonal. The fruit is smaller and seedier than *Citrus latifolia* with a thinner rind but also is more acidic and has a stronger more complex aromatic flavor profile. This lime is also sold and used when green but ripens to a yellow color. It lends its name to key lime pie and is also the lime of choice for margaritas. This lime predominates not only in Mexico but also in many other parts of the world.

Lime Processing

There are three basic alternatives when processing citrus fruit for juice and essential oils: extraction of the essential oil from the whole fruit prior to juice extraction, extraction of the essential oil from the peel following juice extraction, and simultaneous extraction of juice and essential oil.

If the essential oil is extracted before the juice, the juice is less likely to be contaminated by noticeable levels of essential oil, which must be removed to produce juice with an acceptable flavor profile. The quality of the essential oil however, is judged to be higher when separated from the peel following juice extraction.

Lime Essential Oils

Lime essential oils generally fall into two categories: expressed (or cold pressed) and distilled. Expressed oil is washed out of the skin with water once the skin is ruptured due to a grating or pressing process. The oil is separated from the water by centrifuge. Due to limited processing, the flavor of expressed oil is close to that of the natural fruit. It is mid to dark green in color and can be cloudy. In contrast, distilled oil is purified after extraction by distillation producing a clear colorless oil. Many of the components that characterize the flavor of distilled lime are formed during the distillation process, while some volatiles and all non-volatiles are removed. While the distillation process is applied equally to both Tahitian and Key Limes, the majority of commercially available expressed lime is of the Tahitian variety.

Analysis of Lime Oils

A lime oil is a complex mixture consisting of a volatile and non-volatile fraction. The volatile fraction makes up 85-90% of the oil and consists primarily of monoterpene and sesquiterpene hydrocarbons and their oxygenated derivatives along with aldehydes, alcohols, and esters. The non-volatile fraction makes up the additional 10-15% and contains fatty acids, sterols, carotenoids, waxes, coumarins, psoralens, and flavonoids. In a distilled oil, the non-volatile fraction is removed, therefore a comparative analysis considers only the volatile portion of the oil. There are numerous reports of lime oil analysis utilizing a variety of methods and producing a variety of results (*1-3*). A thorough review of citrus oil analysis is available (*4*). The current comparison focuses from the perspective of flavor.

Expressed Lime Oil Comparison

The comparison of expressed Tahitian and Mexican lime oils shows surprisingly little differences. The most prominent differences are the content of β-pinene and γ-terpinene (Fig. 1). The β-pinene content is higher in Mexican or Key lime oil than in the Tahitian or Persian variety. Its taste is described by Mosciano (*5*) as "Fresh, piney and woody, terpy and resinous with slight minty, camphoraceous and spicy nuance." In contrast, γ-terpinene is higher in expressed Tahitian Lime Oil. The taste of γ-terpinene is described as "oily, terpy, musty and citrus like" in taste (*5*).

Distilled Lime Oil Comparison

During the distillation, process the composition of lime oil changes significantly. It is these changes that lead to the characteristic flavor and aroma of distilled lime (Table II). α- and β-pinene concentrations decrease along with neral, geranial, and sabinene leading to the formation of 1,4-cineole, 1,8-cineole, terpinolene, α-terpineol, fenchol, borneol, γ-terpinene, α-terpinene, and *p*-cymene. The primary differences in distilled Mexican or Key Lime and Tahitian lime can be seen in Figure 2. Key Lime is higher in quantities of terpinolene ("woody, terpy, lemon and lime-like with a slight herbal and floral nuance"), α-terpineol ("citrus woody with a lemon and lime nuance"), and β-pinene ("fresh, piney and woody, terpy and resinous with slight minty, camphoraceous with a spicy nuance") (*5*).

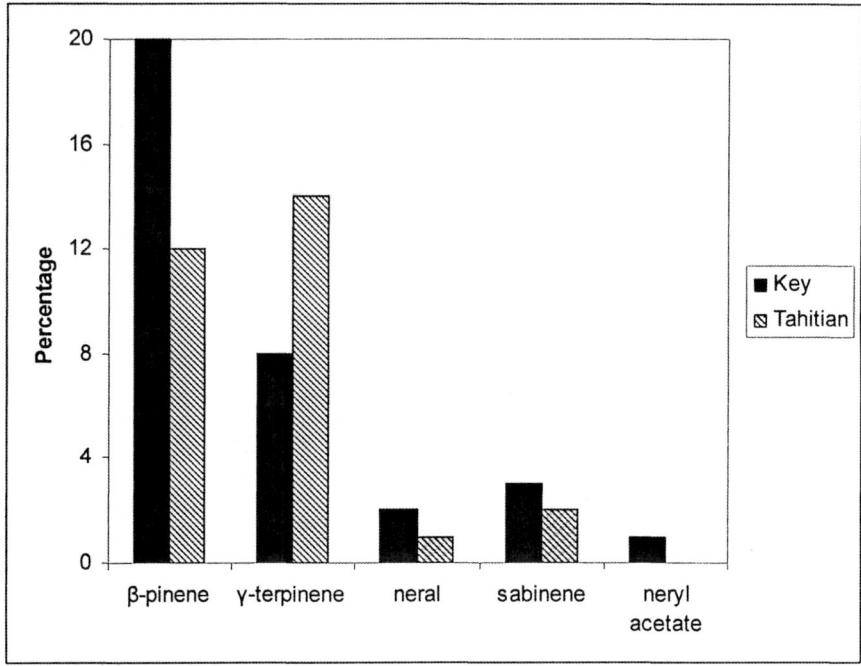

Figure 1. Expressed lime oil composition.

Table II. Lime Oil Comparison

	Distilled lime	*Expressed lime*
Character	Terpeny, fresh, sharp citrus note	Fresh, heavy, sweet, earthy, peel-like, smooth citrus note
Character impact compounds	1,4-cineole, 1,8-cineole, α-terpineol, γ-terpinene	High citral content balanced with β-pinene, γ-terpinene and neryl acetate
Main use	Cola or lime soft drinks	Lemon-lime soft drinks

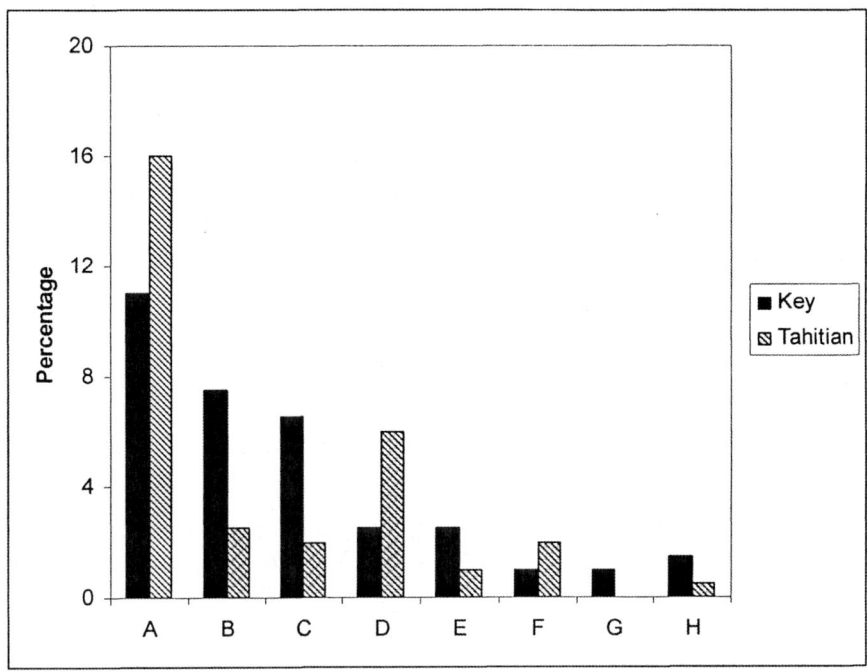

Figure 2. Distilled lime oil comparison. A = γ-terpinene, B = terpinolene, C = α-terpineol, D = α-terpinene, E = β-pinene, F= p-cymene, G = β-bisabolene, H = myrcene.

Tahitian lime has higher levels of γ-terpinene ("terpy, citrus, lime-like, oily, green with a tropical fruity nuance"), α-terpinene ("terpy, woody, piney, citrus lemon and lime with spice and mint nuances"), and p-cymene ("terpy and rancid with slightly woody oxidized citrus notes") (5).

Lime Juice

Lime juice, rind, zest, or essential oils may all be used in different recipes, whether industrially or in the home kitchen. Therefore, both must be considered in a flavor profile comparison. From an aroma chemical standpoint, lime juice is considerably less complex in nature than the essential oil (Table III). Freshly squeezed Mexican lime juice was compared to freshly squeezed Persian lime juice. Organoleptic evaluation (Table IV) correlates well with the analytical results (Table V). Mexican Lime is higher in γ-terpinene (herbaceous-citrus, terpene-like) (6) and β-pinene (resinous-piney odor, turpentine-like) (7).

Table III. Number of Compounds Identified in Lime Juice and Lime Oil

	Juice	Cold Pressed Peel Oil	Distilled Peel Oil
Hydrocarbons	13	36	27
Alcohols	18	32	17
Aldehydes	6	19	7
Ketones	4	8	3
Esters	5	12	2
Sulfur compounds	1	0	0
Ethers	1	2	2
Phenols	0	2	1
Epoxides, pyrans, coumarins	4	12	0
Total	52	123	59

SOURCE: Reference 11.

Table IV. Lime Juice Comparison

	Key Lime Juice	Tahitian Lime Juice
Character	Terpeny, piney, musty notes, balanced lime flavor	Fresh, tart, lime flavor, acid spike, less balanced
Significantly higher in	γ-terpinene, β-pinene, 4-terpineol, bisabolene, geranial	α-terpineol

Conclusion

Natural extracts of lime have different characteristic profiles and differing compositions of aroma chemicals due to lime species, extraction methods, and processing. The appropriate combination of juices, essential oils, and aroma chemicals can be used to simulate typical lime flavor profiles in various applications.

Table V. Major Character Difference Contributors

Aroma Chemical	Description	Higher in
α-Terpineol	Sweet, floral (lilac), lime odor; very sweet taste in dilution; major impact component for lime	Tahitian
γ- Terpinene	Refreshing herbaceous-citrus-like terpene odor	Mexican
β-Pinene	Dry, woody, resinous-piney odor; turpentine-like taste	Mexican
β-Bisabolene	Sweet, balsamic odor, citrus, woody, myrrh, tropical, floral	Mexican
Geranial	Powerful lemon note	Mexican
4-Terpineol	Sweet, earthy-green, musty; slightly peppery woody notes	Mexican

SOURCE: References 6-10.

References

1. Shaw, P.E. *J. Agric. Food Chem.* **1979**, *27*, 246-257.
2. Haro, L.; Faas, W.E. *Perfumer Flavorist* **1985**, *10*(5), 67-72.
3. Guenther, E. *The Essential Oils*; D. Van Nostrand Co.: New York, 1949.
4. *Citrus*; Dugo, G.; Di Giacomo, A., Eds.; Taylor and Francis: London, 2002.
5. Mosciano, G. *Perfumer Flavorist* **2000**, *25*(2), 1-12.
6. Mosciano, G. *Perfumer Flavorist* **1993**, *18*(1), 43-45.

7. Mosciano, G. *Perfumer Flavorist* **1993,** *18*(2), 39-41.
8. Mosciano, G. *Perfumer Flavorist* **1993,** *18*(3), 53-45.
9. Mosciano, G. *Perfumer Flavorist* **1993,** *18*(4), 51-53.
10. Mosciano, G. *Perfumer Flavorist* **1993,** *18*(5), 39-41.
11. *Volatile Compounds in Food: Qualitative and Quantitative Data,* 7th ed.; Maarse, H.; Visscher, C.A.; Willemsens, L.C.; Nijssen, L.M.; Boelens, M.H., Eds.; TNO CIVO Food Analysis Institute: Zeist, The Netherlands, 1994.

Chapter 11

Ethnic Teas and Their Bioactive Components

Amanda M. B. Newell, Sonia Chandra,
and Elvira Gonzalez de Mejia[*]

Department of Food Science and Human Nutrition, University of Illinois
at Urbana-Champaign, 228 ERML, MC–051, 1201 West Gregory Drive,
Urbana, IL 61801

Ethnic teas have been used for centuries to treat diseases. The
objective of this study was to determine the total polyphenol
content, antioxidant, and human DNA anti-topoisomerase II
activities of ethnic teas Ardisia sp., mate (*Ilex paraguariensis*)
and roselle (*Hibiscus sabdariffa* L). Total polyphenols were
determined by using gallic (GA), chlorogenic (CH), and
protocatechuic (PCA) acids as standards. Polyphenols in
ardisia species ranged from 21-72 mg equivalents (eq) GA/g
dried leaves and revealed the presence of epicatechin gallate,
proanthocyanidins, kaempferol, naringenin, and ardisin, with
the highest topoisomerase II catalytic inhibition. Mate tea
contains caffeoyl derivatives (204-364 mg eq. CH/g), which
are significantly different depending on their origin ($P<0.001$).
PCA (33-60 mg eq./g) and anthocyanins were present in
roselle tea. Mate tea presented the highest antioxidant
capacity. Ethnic teas possess compounds with potential health
benefits.

Teas have been utilized for many years for nourishment as well as for their medicinal properties (*1*). Consumption of tea has been associated with the prevention of diseases (*2-4*). The size of the US market for tea is $6.8 billion in 2005, a 9.4 percent increase over the last year. By 2010 it may reach $10 billion. Chemoprevention with dietary substances is an important area of research, which entails the use of non-toxic substances to interfere with carcinogenesis (*5*). Biologically active polyphenols can be found in a variety of plant-derived foods such as fruits, vegetables, nuts, seeds, wine, and tea (*6-9*). Flavonoids, a type of polyphenol, are among the natural compounds found in tea that are gaining attention as having numerous health benefits (*10-12*). More than 4,000 flavonoids have been identified (*13, 14*). These compounds have been shown to exhibit antiproliferative (*4*), apoptotic (*7*), anti-inflammatory, antimicrobial (*15*), antimutagenic, anticarcinogenic, antioxidant (*16*), and therapeutic properties (*6, 17*). A study conducted in Bulgaria found that many of the herbal plants used as tea beverages displayed both a high total polyphenol content as well as antioxidant activity (*9*). Epigallocatechin-3-gallate (EGCG), a component of green tea (GT), has antioxidant and anti-inflammatory activities (*18*). EGCG causes apoptosis to cancer cells, but not to normal cells (*2, 19, 20*). Catalytic inhibition of topoisomerases have demonstrated efficacy in cancer treatments (*21-23*). Several polyphenols, such as EGCG, quercetin, and kaempferol have been reported to have anti-topoisomerase activities (*24*). However, the chemical composition and potential health benefits of ethnic teas still need to be investigated.

Teas made from various ardisia plants (Myrsinace family) have been used for many years in folk medicine for the treatment of liver diseases including liver cancer. Ardisia tea (AT) has been shown to contain flavonoids, which exhibit biological activity (*25*). Rats injected with a known carcinogen and treated with *Ardisia compressa* (AC) did not reveal any signs of carcinogenesis, whereas rats that did not receive the tea treatment developed liver cancer (*26*). Ardisin, a potent flavonoid found in AC has strong anti-topoisomerase activity (*27*). Furthermore, ardisin afforded the highest protection against benomyl-induced oxidation to hepatocytes compared to AC and EGCG in terms of antioxidant capacity (*28*). Embelin, a constituent in ardisia plants, has been used in the treatment of numerous cancers (*29*). Moreover, *A. crenata* extracts demonstrated anti-thrombin activity (80%), compared to controls (*30*). Medicinal properties of ardisia tea including antigenotoxicity, anticytotoxicity, and antioxidant activities have been previously summarized (*25*).

Yerba mate, commonly referred to as mate tea, is an herbal tea prepared from the leaves of *Illex Paraguariensis,* which belongs to the Aquifoliaceae family. Mate tea (MT) grows naturally in Argentina, Uruguay, Brazil, and Paraguay. MT is consumed as a stimulant and tonic beverage (*31*), a central nervous system stimulant, diuretic, and antirheumatic (*32*). The bioactivity of MT is related to its phenolic constituents (*33*). It has been shown that MT has a higher antioxidant capacity than both GT and AC teas (*9, 34*). It also inhibits oral cancer proliferation (*35*). Mate deserves further studies of efficacy and safety (*36*).

Hibiscus sabdariffa (HS), a native plant from West Indies, Nigeria, Sudan, Thailand, Senegal, and Mexico, is used in folk medicine as an antiseptic, astringent, sedative, and tonic. Glew et al. (*37*) have reported the chemical composition of HS indicating the presence of hibiscus acid, anthocyanins (delphinidin and gossypetin or dihidroxyflavone), hibiscin, and protocatechuic acid. HS is the source of a popular red beverage known as "water of jamaica" in Mexico, where it is used as a diuretic. The calyx of the jamaica flower, called karkade in Switzerland, is used as an ingredient in jams, jellies, sauces, and wines. In the West Indies and elsewhere in the tropics the fleshy calyces are used fresh for making roselle wine, jelly, syrup, gelatin, refreshing beverages, pudding, and cakes, and dried roselle is use for tea, jelly, marmalade, ices, ice-cream, sherbets, butter, pies, sauces, tarts, and other desserts. Animal experiments have shown that the consumption of aqueous extracts of HS have antihypertensive and anti-atherosclerotic effects (*38, 39*).

Materials and Methods

Biological Material

Ardisia tea leaves were collected from the greenhouse at the University of Illinois. Fine dried leaves of mate and green teas were obtained from a local market. Hibiscus was obtained from the Mexican company Original Jamaica Real of Veracruz (powder, liquid, and ethanolic preparations).

Chemicals

Trolox (6-hydroxy-2,5,7,8-tetramethylchroman-2-carboxylic acid) and 2,2'-azobis (2-amidinopropane) dihydrochloride (AAPH), were purchased from Aldrich (Milwaukee, WI). All other reagents used were obtained from either Fluka (Milwaukee, WI) or from Sigma Chemical Co. (St. Louis, MO).

Preparation of Ethnic Teas

The aqueous extracts of AT were prepared as described previously (*26*). The freeze-dried solid extracts (SE) or instant teas were kept at -20°C in plastic tubes, sealed with parafilm and protected from light. SE of MT, HS, and GT were prepared following the same standardized procedure as described for ardisia. To standardize the phenolic content, total polyphenol content of SE from every preparation was measured. Values are expressed as milligrams gallic (GA), chlorogenic (CH), or protocatechuic (PCA) acids per milliter and also as per gram dried leaves (DL), depending on the tea.

Polyphenol Content and Characterization

Total polyphenol content of fresh and instant teas were measured as described by the modified Folin-Ciocalteu method (40). Total polyphenol content was expressed as milligram equivalents to the standard used per g of DL. Equations obtained for standard curves were $y = 0.031x - 0.0053$, $r^2 = 0.99$; $y = 0.0052x - 0.006$, $r^2 = 0.99$; and $y = 8E-0.5x^2+0.014x-0.143$, $r^2 = 0.97$; for GA, CH and PCA, respectively. Phenolic compounds were identified using combined analytical techniques including HPLC, LC-MS, TLC, and UV-Vis. A 1050 Hewlett-Packard (Palo Alto, CA) liquid chromatograph with a C_{18} RP guard column and a C_{18} RP Phenomenex Prodigy ODS column were used. A solvent system comprising of solvents A (water/methanol/formic acid, 79.7/20/0.3) and B (methanol/formic acid, 99.7/0.3). The gradient began with 100% A and linearly decreased to 20% A, before returning to 100% A.

Antioxidant Capacity Assays

The ORAC assay (41) was used to assess antioxidant activity by measuring the protection of the tea samples on fluorescent β-phycoerythrin (b-PE) in the presence of free radicals generated by AAPH. The assay was carried out in black-walled 96-well plates (Fisher Scientific, Hanover Park, IL). Immediately after addition of AAPH, the plates were placed in an FL x 800 fluorescence plate reader (Bio-Tek Instruments, Winooski, VT), set with excitation filter 530/25 nm and emission filter 590/35 nm. They were then read every 2 min for 2 hr until reaching a 95% loss of fluorescence. Final fluorescence measurements were expressed relative to the initial reading. Results were calculated based on the differences in the area under the b-PE decay curve, between the blank and a sample and expressed as millimoles of Trolox equivalents [(TEAC)/g dry leaves (DL)]. Trolox is a vitamin E analog that is commonly used to express antioxidant capacity. Trolox (1-4 μM) was used as a standard ($y = 3.35x + 0.42$, $r^2 = 0.98$).

The DPPH (1-diphenyl-2-picrylhydrazyl), a free radical scavenging assay, was also used to evaluate antioxidant capacity (42, 43). The degree of discoloration was indicative of the scavenging ability of the tea samples. The samples were analyzed at 30 min (approximate 50% decrease in initial DPPH by antioxidant) and compared to the standard curve of Trolox (y = 981.14x + 15.55, r^2 = 0.98). Antiradical activity (ARA) of tea samples was calculated as follows: ARA_{sample} = 100 x [1-($sample_{t=n}$ / $control_{t=n}$)], where $_{t=n}$ refers to average absorption value at that time interval. Trolox equivalent antioxidant capacity (TEAC) was determined by the following equation: $TEAC_{sample}$ = ΔA_{sample}/(slope [sample]), where ΔA_{sample} is the decrease in absorbance at 30 min and [sample] is the concentration of the sample (20 mg/mL).

Human Catalytic DNA Anti-Topoisomerase II Assay

Analysis of inhibitory activity against human DNA topoisomerase II was conducted using a topoisomerase drug screening kit (TopoGEN, Inc., Columbus, OH). Electrophoresis was performed in TAE buffer (40 mM Tris-acetate pH 8.0, 1 mM EDTA) using Classic™ CSSU 2025 Electrophoretic Gel System, E-C Apparatus Corporation (Florida, USA) and run at 85 V (2.5 V/cm) for 3.5 hr. Gels were then stained in 0.05 μL/mL ethidium bromide and destained prior to digital imaging using Kodak Image Station 440 CF (Eastman Kodak Company, West New Haven, CT). The band identification was analyzed using Kodak 1D Image Analysis Software version 3.5. The inhibitory activity was measured based on the relative intensity of the supercoiled DNA band in the reaction that contained the tea samples, in comparison to a control. Thus, the higher intensity of this band indicated a higher inhibition on topoisomerase enzyme by the tea compounds.

Results and Discussion

Total Polyphenol Content

Table I presents the total polyphenol content of the various AT. *A. escallonioides* showed the highest total polyphenol content, while *A. japonica* displayed the lowest for both fresh and instant tea samples. Polyphenols in fresh tea ranged from 0.5 ± 0.01 to 1.07 ± 0.01 mg eq GA/mL, while the instant tea spanned 21.4 ± 0.7 to 72.6 ± 1.2 mg eq GA/g DL, for *A. escallonioides* and *A. japonica*, respectively. Freeze-drying the samples to produce instant tea did not alter the total polyphenol content significantly. Instant AC contained 45.5 ± 8.5 mg eq GA/g DL. The total phenolic content of AC was within the range of phenolic contents previously reported (*34*).

Table I. Total Polyphenol Content of Teas Made from Different Ardisia Species[a,b]

Ardisia Plant	Fresh Tea mg GA/g DL	Instant Tea mg GA/g DL
A. escallonioides	1.07 ± 0.01	72.62 ± 1.24
A. elliptica	1.00 ± 0.02	72.31 ± 3.35
A. mamillata	0.91 ± 0.01	59.77 ± 2.45
A. crenata	0.79 ± 0.01	45.47 ± 8.52
A. compressa	1.05 ± 0.01	52.11 ± 0.15
A. japonica	0.50 ± 0.01	21.38 ± 0.70

[a]Average concentration of phenolic compounds. [b]Results from three independent studies. Abbreviations: GA-gallic acid, DL-dried leaves.

Various MT products from different origins were analyzed for total polyphenol content (Table II). Traditional MT only contained mate, while non-traditional contained a variety of other components such as milk, sugar, and artificial flavoring. It was found that traditional MT displayed higher total polyphenol concentrations than the non-traditional teas. It was observed that the total polyphenol content in traditional products ranged from 3.4 to 7.4 mg CH/mL fresh tea, while non-traditional samples ranged from 0.02 to 1.8 mg CH/mL fresh tea. This is probably due to the presence of the additional ingredients. It was also found that MT contains 90-176 mg eq. GA/g DL and 236-490 mg eq CH/ g DL.

Commercially available Hibiscus products were analyzed and compared to fresh tea made from dried HS leaves prepared under culinary conditions. The total polyphenol content is presented in Table III. Health and Healther commercial tea had the highest total phenolic content (58.6 ± 0.1 mg eq. PCA/g DL) while the lowest was found in Tazo Pasion (33.5 ± 0.1 mg eq. PCA/g DL). Fresh tea was found to have 23.9 ± 0.6 mg eq. PCA/DL, lower than that of any of the commercial products.

It was found that the aqueous extract of AC had a different chromatogram profile than MT and GT (Fig. 1). Identification of the compounds present in AC revealed catechins, flavonols, quinone derivatives, isorhamnetin (m/z=317.0), kaempferol (m/z=287), naringenin isomers (m/z=273), and ardisin derivatives. Ardisin is a potent compound showing biological activity. Studies have demonstrated strong anti-topoisomerase activity as well as high protection against benomyl-induced oxidation in hepatocytes with ardisin (26, 27). Other polyphenols that were present in AC include quercetin and rutin. GT is used as a control and both AC and GT had gallic acid and catechins, while MT contained chlorogenic acid and caffeyol derivatives.

Table II. Total Polyphenol Content as mg eq Chlorogenic Acid (CH)/mL and Other Characteristics of Mate Tea Products from Different Origins[a]

Mate Product	Origin	mg eq. CH/mL	Observations
		Traditional Mate Teas	
Cruz de Malta	Buenos Aires, Argentina	3.4 ± 0.7	Fine powder of stem and leaf particles
Amanda Classic	Misiones, Argentina	6.7 ± 0.6	Fine powder of stem and leaf particles
		Non-Traditional Mate Teas	
Lemon Taragui	Corrientes, Argentina	1.8 ± 0.1	Artifical flavor 1.2%, dried lemon particles 2% in 50 g
Adelgamate and herbs	Misiones, Argentina	1.6 ± 0.1	Sea weeds, Sen bush, *Cymtopogon citratus,* GT
CBSe	Santa Fe, Argentina	0.8 ± 0.3	Powder milk, sugar, instant mate, 19 g protein/100 g

[a] Additional mate products were analyzed, but not included in table.

Table III. Concentration of Phenolic Compounds in Commercial Hibiscus Tea Products as Equivalents of Protocatechuic Acid per mL or DL

Hibiscus Teas	mg eq PCA/mL	mg eq. PCA/g DL
Health and Healther	0.72 ± 0.07	58.6 ± 0.07
Bigelow	0.59 ± 0.09	49.1 ± 0.10
Lipton	0.56 ± 0.23	46.8 ± 0.23
Tazo Pasion	0.41 ± 0.06	33.5 ± 0.06
Fresh Tea	0.29 ± 0.06	23.9 ± 0.60

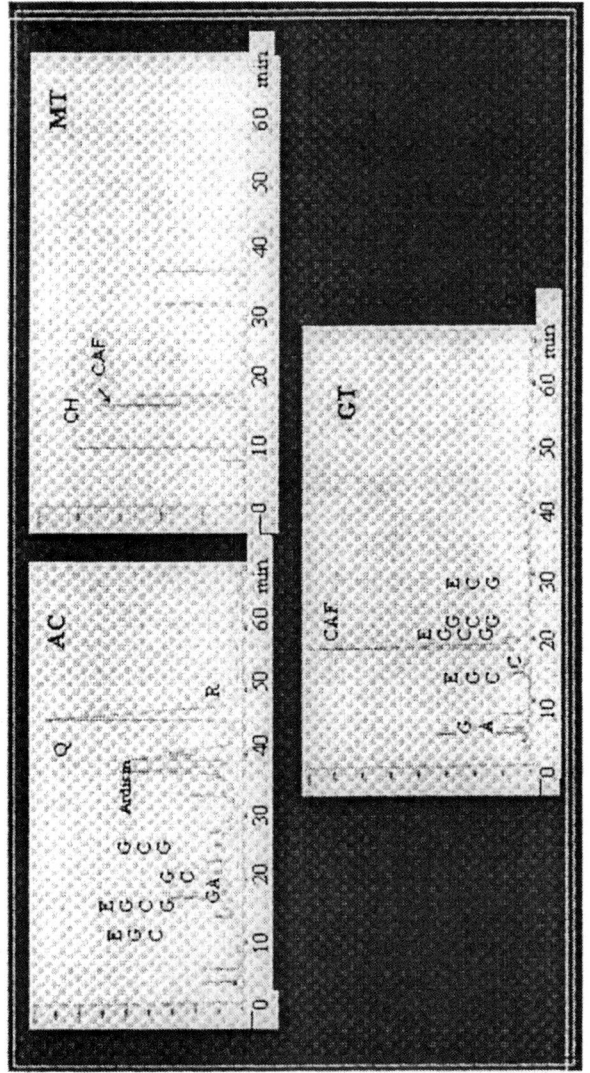

Figure 1. Chromatogram profile of Ardisia compressa, mate, and green teas. Abbreviations: EGC-epigallocatechin, EGCG-epigallocatechin gallate, GA-gallic acid, GCG-gallocatechin gallate, Q-quercetin, R-rutin, CH-chlorogenic acid, CAF- caffeyol derivates, C-catechin, ECG-epicatechin gallate. (Reproduced from reference 34. Copyright 2004 American Chemical Society.)

Antioxidant Activity

In the present study, ORAC and DPPH demonstrated that AT has lower antioxidant capacity than GT and MT. The ORAC assay involves a hydrogen atom transfer, while DPPH involves an electron transfer reaction (*44*). Prior et al. (*45*) reported that ORAC displayed a higher biological relevance over DPPH as well as other popular antioxidant assays because they simulate *in vivo* behavior. The antioxidant capacity of AC, MT, and GT, using the ORAC assay, are summarized in Table IV. MT showed the highest antioxidant capacity per μg GA eq. (13.1 nmol TEAC/μg) ($P < 0.001$), followed by GT (9.1 nmol TEAC/μg) and AC (8.5 nmol TEAC/μg). These results agree with the fact that MT is also a potent inhibitor of nitrosation (*46*). However, when the antioxidant capacity was expressed as μmol TEAC/mL or as μmol TEAC/g DL, GT, with the highest total polyphenol content, showed the highest antioxidant capacity. AC (333.5 μmol TEAC/g DL or 4.9 μmol TEAC/mL) had significantly lower ORAC values than MT (1238.9 μmol TEAC/g DL or 17.4 μmol/mL) and GT (1345.9 μmol TEAC/g DL or 20.1 μmol TEAC/mL). The values obtained in this study agree with the range of ORAC values reported for other teas (235-1526 μmol TEAC/g DL) (*40*).

Table IV. Antioxidant Capacity of Mate, Green and *Ardisia compressa* Teas Using ORAC

Tea Sample	ORAC Value[1,2]		
	nmol TEAC/μg equivalents Gallic Acid	μmol TEAC/mL	μmol TEAC/g DL
Mate	13.1 ± 0.6^a	17.4 ± 0.8^a	1238.9 ± 55.7^a
Green	9.1 ± 0.4^b	20.1 ± 0.9^a	1345.9 ± 60.0^a
Ardisia	8.5 ± 0.2^b	4.9 ± 0.1^b	333.5 ± 8.2^b

[1]Antioxidant capacity relative to 1 μM Trolox. Values are the average of three independent tea preparations. [2]Values with different numbers the same column are statistically different, $P < 0.001$. Abbreviations: TEAC, Trolox equivalent; DL, g dried leaves.

SOURCE: Reproduced from reference 34. Copyright 2004 American Chemical Society.

Antiradical activity (ARA) was reported as the ability of the sample to reduce the free radical DPPH. A higher value corresponds to a higher antiradical activity. It was found that all of the tea samples displayed a relatively high ARA. MT presented the highest antiradical activity. Hibiscus commercial liquid and commercial powder samples exhibited the highest ARA value for the HS, at

81.8% and 80.5%, respectively. The hibiscus ethanolic extract demonstrated the lowest ARA value of all tested samples.

Figure 2 compares the antioxidant activity of a variety of tea samples using the DPPH method. The trolox equivalent antioxidant capacity (TEAC) for AC (36.80 mM TEAC/g SE) was significantly different than the other AT. A comparison of TEAC of a variety of teas, as well as of GA, can be seen in Figure 2b. MT displayed a similar TEAC value as pure GA. It can be observed that GT (40.2 ± 2.8 mM TEAC/g SE) is significantly different than both MT (42.6 ± 2.6 mM TEAC/g SE) and GA (44.4 ± 2.3 mM TEAC/g SE). Fresh HS and one of the commercial hibiscus products extracted with ethanol were found to have the lowest antioxidant capacity. However, when commercial powders and commercial liquids were analyzed, they displayed a TEAC value similar to GT and to some AT (39.8 ± 3.6 and 40.2 ± 2.8 mM TEAC/g SE, respectively).

Topoisomerase II Inhibition

Human DNA anti-topoisomerase II catalytic inhibitory activity by AT is shown in Figure 3. The inhibition of both topoisomerase I and II by AC has been reported (26). The significance of the IC_{50} value is to determine the concentration of tea that will inhibit 50% of the enzyme activity. AC displayed the lowest IC_{50} value (12.3 µg/mL), demonstrating a relatively high anti-topoisomerase II activity. Ramírez et al. (47) reported that the IC_{50} value for AC was 13.8 µg eq. (+) catechin/mL. The IC_{50} values ranged from 12.3 µg/mL (AC) to 389.1 µg/mL (*A. crenata*). AC and *A. japonica* displayed inhibition at lower concentrations, while the four other teas demonstrated little inhibition until the concentration was progressively increased to around 100 µg /mL.

Conclusions

Ethnic teas are a rich source of polyphenols. The flavonoid content depends upon the type of tea as well as the preparation method. The chromatogram of AC shows flavanone and quinone derivatives as the main components, while MT contains mostly caffeyol derivates. HS contains protocatechuic acid and anthocyanidins.

This study revealed that MT contains a higher total polyphenol content and free radical scavenging capacity than both AT and HS. In fact, MT displayed an antioxidant capacity statistically the same as pure GA (20 mg/mL). This study utilized two different methods to assess the antioxidant activity of various teas. Both the ORAC assay and the DPPH method demonstrated that AC has a lower antioxidant capacity than both GT and MT. Although most polyphenols have

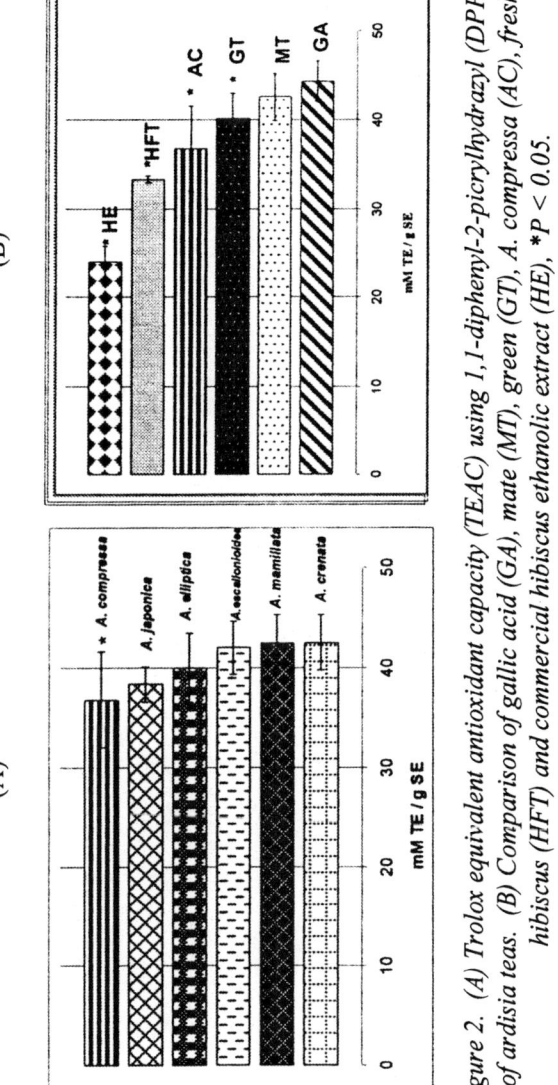

Figure 2. (A) Trolox equivalent antioxidant capacity (TEAC) using 1,1-diphenyl-2-picrylhydrazyl (DPPH) of ardisia teas. (B) Comparison of gallic acid (GA), mate (MT), green (GT), A. compressa (AC), fresh hibiscus (HFT) and commercial hibiscus ethanolic extract (HE), *P < 0.05.

138

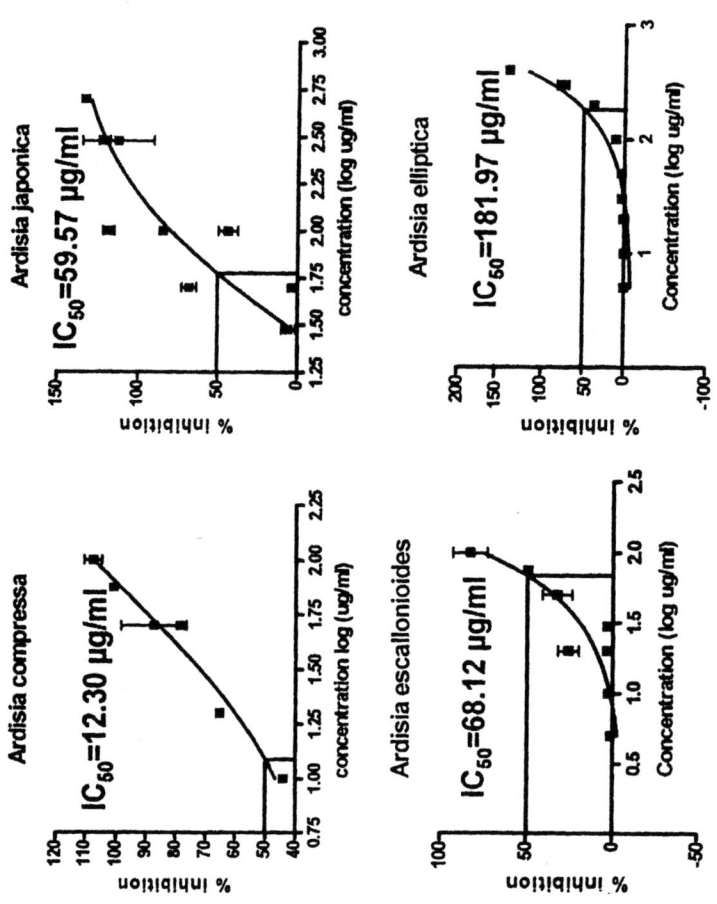

139

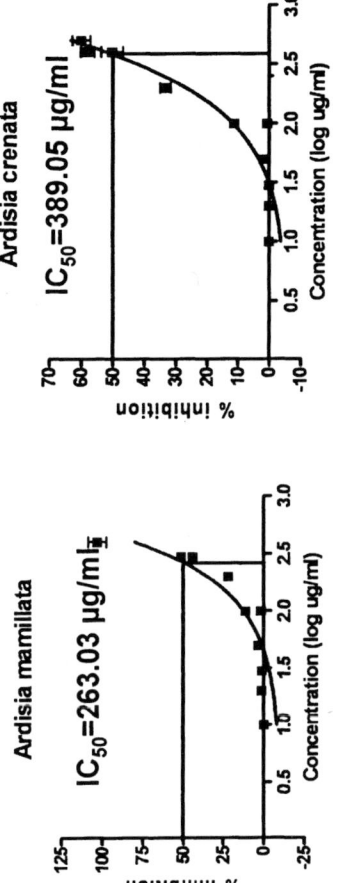

Figure 3. DNA topoisomerase II catalytic enzyme inhibitory activity of ardisia teas. Values are calculated as the percent of inhibition of topoisomerase II by ardisia teas. The concentration is expressed in log values.

140

antioxidant properties, these properties alone may not account for all of their beneficial effects.

Topoisomerase II is the cellular target of AT, mainly acting as a true catalytic inhibitor. AC displayed the best inhibition of topoisomerase II with an IC_{50} of 12.3 µg/mL. Further studies are needed to explore the mechanism of action and the potential of these ethnic teas as chemopreventive and therapeutic agents.

Acknowledgments

Special thanks to Young Soo Song, Laura Kim, Dr. Guadalupe Loarca-Piña, and Dr. Walter Hurley for their collaboration with this study. Also, the authors thank the University of Illinois Research Board for their generous funding to support this research.

References

1. Chen, J.; Tipoe, G. L.; Liong, E. C.; So, H. S.; Leung, K.; Tom, W.; Fung, P. C.; Nanji, A. A. *Am. J. Clin. Nutr.* **2004**, *80*, 742-751.
2. Mukhtar, H.; Ahmad, N. *Am. J. Clin. Nutr.* **2000**, *71*, 1698S-1702S.
3. Kris-Etherton, P. M.; Hecker, K. D.; Bonanome, A.; Coval, S. M.; Binkoski, A. E.; Hilpert, K. F.; Griel, A. E.; Etherton, T. D. *Am. J. Med.* **2002**, *113*, 71S-88S.
4. Crespy, V.; Williamson, G. *J. Nutr.* **2004**, *134*, 3431S-3440S.
5. Bode, A. M.; Dong, Z. *Nutrition* **2004**, *20*, 89-94.
6. Galati, G.; O'Brien, P. J. *Free Radic. Biol. Med.* **2004**, *37*, 287-303.
7. Ramos, S.; Alía, M.; Bravo, L.; Goya, L. *J. Agric. Food. Chem.* **2005**, *53*, 1271-1280.
8. Atoui, A.; Mansouri, A.; Boskou, G.; Kefalas, P. *Food Chem.* **2005**, *89*, 27-36.
9. Ivanova, D.; Gerova, D.; Chervenkov, T.; Yankova, T. *J. Ethnopharmacol.* **2005**, *96*, 145-150.
10. Rietveld, A.; Wiseman, S. *J. Nutr.* **2003**, *133*, 3285S-3292S.
11. Rice-Evans, C. A.; Miller, N. J.; Paganga, G. *Free Radic. Biol. Med.* **1996**, *20*, 933-956.
12. Choudhary, A.; Verma, R. J. *Food Chem. Toxicol.* **2005**, *43*, 99-104.
13. Whitman, S. C.; Kurowska, E. M.; Manthey, J. A.; Daugherty, A. *Atherosclerosis* **2005**, *178*, 25-32.
14. Tereschuk, M. L.; Baigori, M. D.; de Figueroa, L. I.; Abdala L. R. *Methods Mol. Biol.* **2004**, *268*, 317-330.

15. Rauha, J.; Remes, S.; Heinonen, M.; Hopia, A.; Kähkönen, M.; Kujala, T.; Pihlaja, K.; Vuorela, H.; Vuorela, P. *Int. J. Food Microbiol.*. **2000**, *56*, 3-12.

16. Rah, D. K.; Han, D.; Baek, H. S.; Hyon, S.; Park, J. *Toxicol. Lett.* **2005**, *155*, 269-275.

17. Williams, R. J.; Spencer, J. P.; Rice-Evans, C. *Free Radic. Biol. Med.* **2004**, *36*, 838-849.

18. Moyers, S. B., Kumar, N. B. *Nutr. Rev.* **2004**, *62*, 204-211.

19. Chung, J. H.; Han, J. H.; Hwang, E. J.; Seo, J. Y.; Cho, K. H.; Kim, H. K.; Youn, J. I.; Eun, H. C. *FASEB J.* **2003**, *17*, 1913-1915.

20. Kuo, P.; Lin, C. *J. Biomed. Sci.* **2003**, *10*, 219-227.

21. Antony, S.; Arimondo, P.; Sun, J.; Pommier, Y. *Nucleic Acids Res.* **2004**, *32*, 5163-5173.

22. Bal, C.; Baldeyrou, B.; Moz, F.; Lansiaux, A.; Colson, P.; Kraus-Berthier, L.; Léonce, S.; Pierré, A.; Boussard, M.; Rousseau, A.; Wierzbicki, M.; Bailly, C. *Biochem. Pharm.* **2004**, *68*, 1911-1922.

23. Jahnz, M.; Medina, M. A.; Schwille, P. *Chembiochem.* **2005**, *6*, 920-926.

24. Berger, S. J.; Gupta, S.; Belfi, C. A.; Gosky, D. M.; Mukhtar, H. *Biochem. Biophys. Res. Commun.* **2001**, *288*, 101-115.

25. Kobayashi, H.; de Mejía, E. G. *J. Ethnopharmacol.* **2005**, *96*, 347-354.

26. de Mejia, E. G.; Ramirez-Mares, M. V.; Arce-Popoca, E.; Wallig, M.; Villa-Trevino, S. *Food Chem. Toxicol.* **2004**, *42*, 509-516.

27. de Mejía, E. G.; Ramírez-Mares, M. V.; Nair, M. G. *J. Agric. Food Chem.* **2002**, *50*, 7714-7719.

28. Ramirez-Mares, M. V.; de Mejia, E. G. *Food Chem. Toxicol.* **2003**, *41*, 1527-1535.

29. Nikolovska-Coleska, Z.; Xu, L.; Hu, Z.; Tomita, Y.; Li, P.; Roller, P. P.; Wang, R.; Fang, X.; Guo, R.; Zhang, M.; Lippman, M. E.; Yang, D.; Wang, S. *J. Med. Chem.* **2004**, *47*, 2430-2440.

30. Chistokhodova, N.; Nguyen, C.; Calvino, T.; Kachirskaia, I.; Cunningham, G.; Miles, D. H. *J. Ethnopharmacol.* **2002**, *81*, 277-280.

31. Bracesco, N.; Dell, M.; Rocha, A.; Behtash, S.; Menini, T.; Gugliucci, A.; Nunes, E. *J. Altern. Complement. Med.* **2003**, *9*, 379-387.

32. Tapiero, H.; Tew, T. D.; Nguyen, G.; Mathe, G. *Biomed Pharmacother.* **2002**, *56*, 200-207.

33. Filip, R.; Lopez, P.; Giberti, G.; Coussio, J.; Ferraro, G. *Fitoterapia* **2001**, *72*, 774-778.

34. Chandra, S., de Mejia, E. G. *J. Agric. Food Chem.* **2004**, *52*, 3583-3589.

35. de Mejia, E. G.; Song, Y. S.; Ramirez-Mares, M. V.; Kobayashi, H. *J. Agric. Food Chem.* **2005**, *53*, 1966-1973.

36. Goldenberg, D.; Golz, A.; Joachims, H. Z. *Head Neck* **2003**, *25*, 595-601.

37. Sena, L. P.; Vander Jagt, D. J.; Rivera, C.; Tsin, A. T. C.; Muhamadu, I.; Mahamadou, O.; Millson, M.; Pastuszyn, A.; Glew, R. H. *Plant Foods Hum. Nutr.* **1998**, *52*, 17-30.

38. Odigie, I. P.; Ettarh, R. R.; Adigun, S. A. *J. Ethnopharmacol.* **2003**, *86*, 181–185.

39. Chen, C. C.; Hsu, J. D.; Wang, S. F.; Chiang, H. C.; Yang, M. Y.; Kao, E. S.; Ho, Y.C.; Wang, C. J. *J. Agric. Food Chem.* **2003**, *51*, 5472–5477.

40. Nurmi, K.; Ossipov, V.; Haukioja, E.; Pihlaja, K. *J. Chem. Ecol.* **1996**, *22*, 2023-2040.

41. Prior, R. L; Cao, G. *Proc. Soc. Exp. Biol. Med.* **1999**, *220*, 255-261.

42. Fukumoto, L. R.; Mazza, G. *J. Agric. Food Chem.* **2000**, *48*, 3597-3604.

43. Cardador-Martínez, A.; Loarca-Piña, G.; Oomah, D. B. *J. Agric. Food Chem.* **2002**, *50*, 6975-6980.

44. Huang, D.; Ou, B.; Prior, R. L. *J. Agric. Food Chem.* **2005**, *53*, 1841-1856.

45. Prior, R. L.; Wu, X.; Schaich, K. *J. Agric. Food Chem.* **2005**, *53*, 4290-4302.

46. Bixby, M.; Spieler, L.; Menini, T.; Gugliucci, A. *Life Sci.* **2005**, *77*, 345-358.

47. Ramirez-Mares, M. V.; Chandra, S.; de Mejia, E. G. *Mutat. Res.* **2004**, *554*, 53-65.

Chapter 12

Identification of Characteristic Aroma Components of Mate (*Ilex paraguariensis*) Tea

Patricio R. Lozano, Keith R. Cadwallader,
and Elvira González de Mejia

Department of Food Science and Human Nutrition, University of Illinois
at Urbana-Champaign, 1302 West Pennsylvania Avenue, Urbana, IL 61801

Mate (*Ilex paraguariensis*), native to South America, has been used for centuries by the South American Indians as a source for a tonic and a stimulating drink, and recent discoveries have associated its phenolic content with the inhibition of tumor growth. Despite the health promoting properties of Mate tea, its characteristic flavor is still not well-characterized. Three types of mate tea imported from Argentina were selected based on levels of polyphenolic compounds, agronomic factors, and flavor characteristics (strong, medium, and weak flavor) determined by habitual Mate tea drinkers. Volatiles from hot water-infusions (brews) prepared from dried tea leaves were isolated by dynamic headspace analysis (DHA), solvent-assisted flavor evaporation-solvent extraction (SAFE-SE), and column adsorption extraction. Aroma-active components were identified by gas chromatography-olfactometry (GCO) and GC-MS. Predominant aroma components of Mate tea included geraniol, β-damascenone, 2-methoxyphenol, linalool, β-ionone, eugenol, 2-acetyl-1-pyrroline, (*E,Z*) 2,6-nonadienal, and geranial. Aroma components of Mate tea are produced by lipoxygenase action and degradation reactions occurring during manufacture of the dried tea leaves.

143

Introduction

Mate tea is a beverage prepared by the infusion of dried green Mate leaves and is traditionally consumed in the southeastern part of South America including Paraguay, Uruguay, Brazil, Argentina, and Bolivia. Mate is well known for its high content of alkaloids such as caffeine, which can have pharmaceutical applications. Caffeine intake due to Mate tea consumption by far exceeds intakes recorded in the literature for other beverages containing this alkaloid (*1*). Extraction of caffeine from Mate can be accomplished by use of supercritical CO_2, which gives the highest yield (*2*). Antioxidative properties have been found in Mate tea and its polyphenolic compounds have been associated with inhibition of tumor growth (*3*) and nitrosative stress (*4*). Despite the importance of Mate tea leaves in the social and economic context of South America, the literature is scarce in relation to its characteristic flavor.

Mate tea has a characteristic mature flavor, which is somewhat sweet, sour leaf-like and cigarette-like. This flavor is similar to that obtained from *Camellia sinensis* tea (*5*). Volatile components of Argentinean green Mate and Brazilian roasted Mate were investigated by Kawakami and Kobayashi (*6*). A total of 196 compounds were identified as volatile constituents of Mate in that study; however, no sensory (olfactometry) assessment was carried out. Therefore, the aroma significance of these aroma compounds has not been fully addressed.

The aim of the present study was to identify and characterize the major aroma components of mate tea using novel extraction techniques in combination with GCO and GC-MS.

Experimental Procedures

Materials

Three samples of commercial Mate tea imported from Argentina. Cruz de Malta low ash, Kraus Organic Classic Mate, and Taragui low ash were selected for this study based on the levels of polyphenolic compounds (*3*), agronomic conditions (*7*) and flavor characteristics suggested by traditional Mate drinkers, who classified the samples as having low, medium or high flavor intensity, respectively. The samples were sealed in plastic bags and frozen at -80°C until analysis.

Preparation of Mate Tea Infusion

Sample brewing process was carried out following the protocol described by Chandra and Gonzalez de Mejia (*3*). Each tea (2.7 g) was mixed with 250 mL of boiling water at 98°C and held/stirred for 10 min. The tea extract was then filtered using Whatman paper #2.

Dynamic Headspace Dilution Analysis (DHDA)

Ten mL of tea brew was placed in a glass three-neck purge and trap vessel (SIS, CA). A desorption tube containing 200 mg of Tenax TA (Supelco, Bellefonte, PA) was attached to the vessel. The vessel was kept at 50°C for 10 min to allow equilibration of the headspace of the tea and then nitrogen at a constant flow of 50 mL/min was used to purge the headspace volatiles onto the Tenax trap. Serial dilutions were accomplished by varying the purging time from 25, 5, or 1 min, corresponding to flavor dilution factors of 1, 5, and 25 respectively (*8*). Prior to thermal desorption the trap was dry purged for 17 min (100 mL/min nitrogen flow) to remove excess moisture. A thermal desorption system (TDS2, Gerstel, Germany) was employed to desorb (220°C) and transfer the volatiles from the Tenax trap into a CIS 4 inlet (Gerstel) where they were cryofocused at -150°C. Splitless injection was made by ramping the CIS 4 temperature to 260°C at 12°C/s. GCO was performed on a 6890 GC (Agilent Technologies, Palo Alto, CA) equipped with an OD2 olfactometry port (Gerstel) and Stabilwax-DA (Restek, Bellefonte, PA) or DB5ms (J&W Scientific, Folson, CA) capillary column (15 m x 0.32 mm i.d. x 0.5 μm film). Analyses were carried out in duplicate.

Solvent Assisted Flavor Evaporation – Solvent Extraction (SAFE-SE)

A modified solvent assisted flavor evaporation (SAFE) apparatus (*9*) was used for the extraction of the main aroma components of Mate tea brew. SAFE-SE was selected because it allows for a representative aroma extract to be prepared without the formation of artifacts. Mate tea brew (800 mL) was spiked with 10 μL of internal standard solution (50 mg each of 2-undecanol, 6-undecanone, *tert*-butylbenzene, and 2-ethyl butyric acid dissolved in 10 mL of methanol) and kept in an Erlenmeyer flask covered with aluminum foil to prevent any light affects on the chemical composition of the tea during the extraction time. The SAFE extraction was conducted under the following conditions: vacuum at 10^{-5} Torr, 45°C SAFE head temperature, and 40°C water

146

bath (sample flask) temperature. The receiving traps were kept in liquid nitrogen temperature (-196°C) throughout the distillation. The distillate was thawed at room temperature and extracted with ether (3 x 50 mL). The ether extract was separated into acidic, neutral and basic fractions (10). Each fraction was dried over 2 g of sodium sulfate and then concentrated to 200 μL under a gentle stream of nitrogen.

Adsorptive Column Extraction

Mate tea brew (800 mL) was passed through a column packed with 10 g of Porapak Q (50- 80 mesh, Waters Co., Milford, MA) using the method described by Kumazawa and Masuda (11) and Maneerat et al. (12). The column was washed with 100 mL deodorized-deionized water and the adsorbed compounds were then eluted with 100 mL of methylene chloride. The elute was separated into neutral, acidic and basic fractions and each fraction was concentrated to 200 μL as earlier described.

Aroma Extract Dilution Analysis (AEDA)

GCO was conducted using an Agilent 6890 GC equipped with a DATU sniffing port (Geneva, NY) and flame ionization detector. Separations were performed using polar (Stabilwax DA) and nonpolar (DB5ms) (15 m x 0.25 mm i.d. x 0.25 μm film) columns. Serial dilutions (1:3) were prepared from each aroma fraction from SAFE-SE and column adsorptive extraction. Two μL of each dilution was injected in the cool on-colum mode. Oven temperature was 35°C for 5 min, then ramped to 225°C at 4°C/min, with a final holding time of 20 min. The carrier gas was He at a constant flow of 1 mL/min.

Gas Chromatography-Mass Spectrometry (GC-MS)

The GC-MS system consisted of an Agilent 6890 GC/5973 MSD equipped with a cool on-column injector. Separations were performed on a Stabilwax-DA column (30 m x 0.25 mm i.d. x 0.5 μm; Restek). GC conditions were the same as for AEDA except that only 1 μL of each extract was injected. MSD conditions were as follows: 280°C interface temperature, ionization voltage was set at 200 V above standard tune value and mass range was 35 – 300 a.m.u.

Compounds were positively identified by comparing their retention indices (on two columns) and mass spectra with authentic reference compounds. Tentative indentifications were based on matching RIs with literature values or by comparing their mass spectra to a database (Wiley 138K Mass Spectral Database, Wiley and Sons, 1990).

Results and Discussion

Aroma Components Isolated by SAFE-SE

Forty-two, 23, and 45 aroma compounds were detected in Cruz de Malta, Kraus Organic, and Taragui Mate tea infusions, respectively. However, some of the aroma compounds that were detected at the sniffing port could not be identified by GC-MS due to their low abundance. Table I shows only the major aroma compounds identified in these three Mate extracts.

Although a higher number of compounds were detected in the Taragui Mate tea extract by using SAFE-SE, all of the Mate teas seemed to show similar aroma-active compounds, which provide the chracterisitic herbal/minty, grassy/hay and burnt/tabbaco aroma to the tea.

Cruz de Malta tea, which was characterized by a smoky and tobbaco note, showed the lowest intensity in its aroma compounds. β-ionone, eugenol, 2-methoxy-4-vinylphenol (p-vinylguaiacol), geraniol, 2-methoxy phenol (guaiacol), geranial, and 1,8-cineole were the major aroma contributors in this extract. From these compounds 2-methoxyphenol and 2-methoxy-4-vinylphenol have been reported previously in other types of teas as important contributors of roasted/smoky aroma notes (11).

Taragui tea, which was classified by habitual mate consumers as having the strongest flavor among the three teas analyzed in this study, possessed more intense grassy and floral aroma notes besides the burnt and hay aroma discussed above. It appears that the presence of hydroperoxide breakdown products such as (E)-2-octenal (peanut), (Z)-3-hexenal (grassy), 1-octen-3-ol (mushroom), and (E,Z)-2,6-nonadienal (cucumber) contribute to the grassy or green flavor in this type of Mate tea. These compounds were shown to be formed as a result lipoxygenase activity due to mild temperature treatment employed to inhibit leaf enzymes. In addition, geraniol (floral), geranial (fruity), 1,8-cineole (minty), linalool (floral), and eugenol (spicy) contribute to the higher herbal/minty character of the Taragui extract. These terpene alcohols are formed via hydrolysis of glycoside precursors (13).

The aroma extract of Kraus-Organic mate tea contained a lower number of aroma compounds, generally at weaker intensities, than the other two tea extracts. This finding could be due to the different agronomic conditions applied to these leaves, which has been shown to influence the amount of aroma

Table I. Predominant Odorants Determined by SAFE-SE and AEDA of Hot-Water Infusions of Three Types of Mate Tea

Compound	RI^a	Odor Description	Fr^b	FD-factor^c		
				$Cruz^d$	Org^e	$Tara^f$
Geraniol [A]	1864	Floral	N	9	243	729
Guaiacol [A]	1853	Smoky, medicine	N	3	27	243
1-Octen-3-ol[B]	1438	Mushroom	N	- -	3	27
β-Damascenone [A]	1819	Cooked apple	N	3	729	81
δ-Octalactone [A]	1972	Fruity, floral	N	9	9	81
Skatole[B]	2486	Urine, mothballs	N	9	81	81
(Z)-3-Hexenal[B]	1114	Green, cut-leaf	N	3	27	27
1,8-Cineole[A]	1171	Minty, eucalyptus	N	9	3	27
2-Acetyl-1-pyrroline[B]	1319	Roasty, popcorn	B	3	27	27
Methional[B]	1465	Cooked potato	N	<3	3	27
Linalool [A]	1534	Floral	N	- -	1	27
(E)-2-Nonenal[B]	1544	Hay, fatty	N	- -	27	27
(E,Z)-2,6-Nonadienal[B]	1582	Cucumber	N	3	9	27
Butanoic acid [A]	1612	Sweaty, cheesy	A	9	9	27
2-Acetylthiazole[B]	1632	Roasty, popcorn	B	3	3	27
β-Ionone[A]	1911	Peachy, floral	N	27	3	27
γ-Nonalactone[A]	2017	Coconut, sweet	N	3	27	27
p-Cresol[A]	2086	Phenolic, animal	N	3	3	27
Eugenol[A]	2158	Floral, spicy	N	27	81	27
p-Vinyl guaiacol[B]	2191	Cloves, spicy	N	27	81	27
2,3-Butanedione[B]	975	Buttery, creamy	N	3	9	9
1-Penten-3-one[A]	1047	Plastic, rancid	N	3	27	9
(Z)-4-Heptenal[B]	1197	Nutty, rancid	N	<3	3	9
(E)-2-Octenal [A]	1426	Nutty, fatty	N	- -	3	9
(E)-2-Decenal[A]	1663	Green, pungent	N	3	9	9
Pentanoic acid [A]	1716	Sweaty, cheese	A	- -	9	9
Citronellol[A]	1760	Herbal	N	9	9	9
Geranial [A]	1766	Fruity, floral	N	9	27	9
Hexanoic acid [A]	1876	Sweat, ,body odor	A	- -	- -	9
Furaneol[B]	2055	Burnt sugar	A	<3	9	9
Wine lactone[B]	2219	Plastic	N	3	9	9
Unknown	2501	Vanilla	A	3	9	9
2-3-methylbutanal[A]	924	chocolate, malt	N	3	9	3
2,3-Pentanedione[B]	1054	Buttery, creamy	N	9	27	3
1-Hexen-3-one [A]	1086	Plastic, metallic	N	<3	3	<3
(E,E)-2,4-Hexadienal[B]	1444	Fatty, metallic	N	3	3	<3

[A]Compound positively identified (RI, odor, MS). [B]Compound tentatively identified (RI, odor). [a]Retention index on Stabilwax-DA column. [b]Fraction containing odorant. [c]Flavor dilution factor. [d]Cruz de Malta tea. [e]Kraus Organic tea. [f]Taragui tea. - -, not detected.

compounds formed (7). The intensity of linalool in Kraus-Organic was lower than in Taragui, but higher than in Cruz de Malta. Linalool and other terpene alcohols have been found in *Camelia sinensis* tea and pine tree extracts (*14*), and they were associated with an extended aging process and sweeter flavor characteristics.

Maillard reaction products such as 2,3-pentanedione, 2,3-butanedione, methional, and 2-acetyl-1-pyrroline as well as carotenoid thermal degradation products like β-damascenone seemed to contribute to the sweet/buttery aroma of all three Mate tea brews. β-Damascenone and methional have been identified as contributors of the aroma of green tea and were also reported as having higher intensities in black tea infusions that have undergone heat processing (*11*).

The similar main aroma compounds for the three Mate teas suggests that these compounds are mainly responsible for flavor differences detected by the habitual Mate drinkers. Therefore, further characterization by other techniques was conducted to corroborate the intensity of these compounds as well as look for other compounds not recovered by SAFE-SE. The techniques chosen were dynamic headspace analysis and column adsorption extraction. Taragui Mate tea was chosen as being most representative of the typical Mate tea and it also contained a high number and intensity of aroma-active compounds in comparison to the other two teas.

Aroma Components Isolated by Adsorptive Column Extraction

The majority of the compounds identified by the AEDA of adsorptive column extracts (Table II) agreed well with the results of AEDA/SAFE-SE for the Taragui mate tea. Similar to the results of SAFE-SE, floral and roasted notes were predominant in the aroma profile of the tea. Geranial, β-damascenone, geraniol and linalool showed the highest dilution factors. Morever, compounds such as 2-acetyl-2-thiazoline, guaiacol, β-ionone, *p*-cresol eugenol, and *p*-vinyl guaiacol made a moderate contribution to the tea aroma.

Aroma Components Isolated by DHDA

Results of the analysis of the intermediate and highly volatile aroma components by DHDA of Taragui Mate tea are shown in Table III. The main contributors of the characteristic headspace aroma of Mate tea were mainly floral and roasted aromas. Predominant aroma components included geraniol, linalool, β-ionone, and eugenol. Compounds such as guaiacol, 2-acetyl-1-pyrroline, methional, *(E)*-2-octenal, and *(Z)*-4-heptenal provide a smoky/roasted and plant-like aroma notes to the aroma profile of the tea. These compounds most likely are formed during the drying steps of the manufacturing process of the tea leaves and through lypoxygenase activity. In addition to these main contributors of the aroma, complementary compounds such as 2,3-butanedione, 1-penten-3-one, γ-nonalactone, and β-damascenone, among others, impart buttery, burnt sugar and sweet aroma notes to round out the main floral/smoky aroma of the beverage.

Table II. Predominant Odorants Determined by Adsorptive Column Extraction - AEDA of Hot-Water Infusions of Taragui Mate Tea

Compound	RI^a	Odor Description	$Fr.^b$	$FD\text{-}factor^c$
Geranial[A]	1771	Fruity, floral	N	243
β-Damascenone[A]	1819	Cooked apple	N	243
Geraniol[A]	1844	Floral	N	243
Linalool [A]	1555	Floral, honeysuckle	N	81
2-Acetyl-2-thiazoline[B]	1750	Roasty, popcorn	B	81
Guaiacol [A]	1853	Smoky	N	81
β-Ionone [A]	1935	Floral	N	81
p-Cresol[A]	2083	Dung, horse stable	N	81
Eugenol[A]	2157	Cloves, spicy	N	81
p-Vinyl guaiacol[B]	2186	Cloves, smoky	N	81
Unknown	2502	Vanilla	A	81
2-Acetyl-1-pyrroline[B]	1337	Roasty, popcorn	B	27
(E,Z)-2,6-Nonadienal[A]	1492	Cucumber	N	27
2-Acetylthiazole[B]	1634	Roasty	N	27
γ-Nonalactone[A]	2062	Coconut, sweet	N	27
Skatole[B]	2415	Urine, mothballs	B	27
Butanoic acid[A]	1591	Sweaty, cheesy	A	9
Hexanoic acid[A]	1876	Sweaty, body odor	A	9
1-Penten-3-one[B]	1080	Plastic	N	3
1,8-Cineole[A]	1199	Minty	N	3
1-Octen-3-one[B]	1304	Mushroom	N	3
(E)-2-Octenal [A]	1402	Rancid, raw peanut	N	3
Methional[B]	1465	Cooked potato	N	3
Maltol[A]	1932	Burnt sugar	A	3
o-Cresol[A]	2009	Phenolic, medicine	N	3

A, B, a, b, cRefer to footnotes in Table I.

Table III. Predominant Odorants Determined by DHDA of Hot-Water
Infusions of Taragui Mate Tea

Compound	RI[a] FFAP	DB5	Odor Description	FD-factor[b]
2,3-Butanedione	964	<600	Butter, creamy	25
1-Penten-3-one	1027	683	Plastic, water bottle	25
(Z)-4-Heptenal	1246	914	Rancid, crabby	25
1-Octen-3-one	1306	980	Mushroom	25
(E)-2-Octenal	1436	1059	Raw peanut	25
(E)-2-Nonenal	1509	- -	Hay, stale	25
Linalool	1537	1103	Floral, honeysuckle	25
Geraniol	1842	1256	Floral, tea	25
Guaiacol	1857	2089	Smoky	25
β-Ionone	1966	1491	Stale, hay, floral	25
γ-Nonalactone	2018	1363	Coconut, floral	25
Eugenol	2145	- -	Cloves, spicy	25
Hexanal	1056	813	Green, cut-grass	5
1-Hexen-3-one	1067	775	Plastic, water bottle	5
Octanal	1294	1021	Orange oil	5
2-Acetyl-1-pyrroline	1346	929	Roasty, popcorn	5
(Z)-1,5-Octadien-3-one	1383	988	Metallic	5
1-Octen-3-ol	1417	992	Mushroom	5
Methional	1464	912	Cooked potato	5
(Z)-2-Nonenal	1502	1148	Melon, hay	5
(E,Z)-2,6-Nonadienal	1589	1170	Cucumber	5
Butanoic acid	1616	821	Sweaty, cheesy	5
(E)-2-Decenal	1652	1336	Green, pungent	5
Citronellol	1727	1231	Fruity	5
Geranial	1756	- -	Citrus	5
β-Damascenone	1811	1389	Applesauce	5
unknown	2327	- -	Floral	5

[a]Retention index on FFAP (Stabilwax-DA column) and DB5 (DB-5ms) columns. [b]Flavor dilution factor.

Conclusions

The aroma of Mate tea can be described as a mixture of several aroma-active compounds that impart a floral, minty, smoky/tobacco, and sweet/buttery aromas. The herbal and minty attributes of the mate extracts seemed to be associated with terpene alcohols such as geraniol, geranial, β-ionone, linalool, and eugenol which have been isolated from other green teas (15). The smoky, tobacco, burnt, earthy, and hay aromas of Mate tea could be associated with the

152

presence of lipid oxidation and Maillard reaction products that are formed during the drying and aging process of the Mate leaves (*16*). Further research would be required to relate the potency of these compounds with sensory perception.

References

1. Mazaffera, P. *J. Food Chem.* **1997**, *60*, 67-71.
2. Saldaña, M. D.; Zetzl, C.; Mohamed, R. S.; Brunner G. *J. Agric. Food Chem.* **2002**, *50*, 4820-4826
3. Chandra, S.; Gonzalez de Mejia, E. *J. Agric. Food. Chem.* **2004**, *52*, 3583-3589.
4. Bixby, M.; Spieler, L.; Menini, T.; Gugliucci, A. *Life Sci.* **2005**, *77*, 345-358.
5. Yamanishi, T.; Nose, M.; Nakatani, Y. *Agric. Biol. Chem.* **1970**, *34*, 599-608.
6. Kawakami, M.; Kobayashi, A. *J. Agric. Food Chem.* **1991**, *39*, 1275-1279.
7. Aquino Esmelindro, A.; Dos Santos Girardi, J.; Mossi, A.; Assis Jacques, R. and Dariva, C. *J. Agric. Food. Chem.* **2004**, *52*, 1990-1995.
8. Cadwallader, K.R.; Baek, H.H. Aroma-impact components in cooked tailmeat of freshwater crayfish (*Procambarus clarkii*). In *Food Flavors: Formation, Analysis and Packaging Influences;* Contis, E.T., Ho, C.T., Mussinan, C.J., Parliment, T., Shahidi, F., Spanier, A.M., Eds.; Elsevier: New York, 1998, pp 271-278.
9. Engel, W.; Bahr, W.; Schieberle, P. *Eur. Food Res. Technol.* **1999**, *209*, 237-241.
10. Benitez, D. Effect of Presoak Treatment and Sodium Bicarbonate Addition during Processing on Volatile Composition of Soymilk. M.Sc. Thesis, University of Illinois, Urbana-Champaign, IL, 2003.
11. Kumazawa, K.; Masuda, H. *J. Agric. Food Chem.* **2001**, *49*, 3304-3309.
12. Maneerat, C.; Hayata, Y.; Kozuka, H.; Sakamoto, K., Osajima, Y. *J. Agric. Food Chem.* **2002**, *50*, 3401-3404.
13. Yano, M.; Okada, K; Kubota, K; Kobayashi, A. *Agric. Biol. Chem.* **1990**, *54*, 1023-1028.
14. Kim, K-Y.; Chung, H-J. *J. Agric. Food Chem.* **2000**, *48*, 1269-1272.
15. Takeo, T. *Phytochemistry* **1981**, *20*, 2145-2147.
16. Yamanishi, T.; Kawakami, M.; Kobayashi, A.; Hamada, T.; Musalam, Y. Thermal Generation of Aroma Compounds from Tea and Tea Constituents. In *Thermal Generation of Aromas*; ACS Symposium Series 409; Parliment, T.H.; McGorrin, R.J.; Ho, C.T., Eds.; American Chemical Society: Washington, DC, 1989, pp. 310-319.

Chapter 13

Latest Advances in the Chemical and Flavor Characterization of Mexican Distilled Beverages: Tequila, Mezcal, Bacanora, and Sotol

Belinda Vallejo-Córdoba and Aarón F. González-Córdova

Centro de Investigación en Alimentación y Desarrollo, A.C. (CIAD), Apartado Postal 1735, Hermosillo, Sonora, Mexico 83000

Differences among Tequila, Mezcal, Bacanora, and Sotol are described, with special emphasis on volatile characterization and aroma discrimination based on an electronic nose. Direct injection gas chromatography and solid phase microextraction gas chromatography mass spectrometry (SPME-GC/MS) were reported for the characterization of major and minor volatiles. Alcohols, esters, ketones, and aldehydes were identified in the four spirits. The most abundant volatiles after the alcohols were ethyl esters. Quantitative differences among Tequila types and spirits were reported for these volatiles. A micellar electrokinetic capillary electrophoresis (MEEKC) method was presented for the quantitative determination of furfural in the spirits. Furfural concentrations were within permitted limits according to Mexican regulations. Aroma discrimination by an electronic nose allowed differentiation among Tequila types and Tequila, Mezcal, and Bacanora.

153

The diverse varieties of agave plants found throughout Mexico have been used since time immemorial to make different alcoholic beverages. Due to numerous growing conditions affecting the characteristics of agave, the product obtained in each region is completely different from that found in any other region *(1)*. Tequila manufacture begins with the cooking of *Agave tequilana* Weber azul plant heads or "piñas" obtained after removing the fleshy leaves of the agave. Cooking these heads provides the juices, which are transformed after fermentation into Tequila *(2)*. Thus, these piñas are the raw material that possess the source of carbohydrates (mainly inulin), which by the fermentation process are transformed into ethanol. Similarly, other agave varieties from different regions in Mexico are used to produce Mezcal (*A. angustifolia Haw, A. esperrima jacobi, A. weberi cela, A. potatorum zucc, A. salmiana otto*) *(3)*, Bacanora (*A. angustifolia*) *(4)* and Sotol (*Dasylirion spp*) *(5)*. Although the manufacture of Tequila, Mezcal, Bacanora, and Sotol basically consists of cooking, millling, fermentation, and distillation, only Tequila is produced with high technology. The manufacture of Mezcal and Sotol are partly industrialized, while Bacanora continues to be artisanal. Tequila, Bacanora, and Mezcal may be produced from 100% agave *(2-5)*; however, Tequila and Sotol may contain up to 49% sugars different from agave juices *(2, 5)*, while Mezcal and Bacanora may contain only up to 20% *(3, 4)*.

Agaves are cultivated in arid and semi-arid regions worldwide. The genus contains 140 species, which constitute the family Agavaceae *(1)*. However, without any doubt, the creation of these aromatic spirituous drinks from specific agave plants has its origin in Mexico during the prehispanic period. The most ancient information that reveals the existance of agave and its different uses goes back to the era before the Spaniards, and appears in the *Nahuatl* codex. The same codex mentions that certain tribes had learned to cook agave plants in an underground hole and used them as food to compensate for the lack of water in desert lands. Also, these tribes discovered that cooked agave soaked in water could ferment, producing a very appreciated beverage. This primitive and rudimentary method was used for centuries to produce beverages from agave, considered a sacred plant possessing divine properties. Upon the arrival of the Spaniards in Mexico in 1519, Pulque (fermented agave juices), was the only alcoholic drink available. But it was not until then, when the Spaniards brought their knowledge of distillation techniques, that these spirits, also called "agave wine" or "mezcal wine" from which Tequila emerged by the late 1800's, took their present form *(6)*.

The first Tequila production process with a commercial purpose was established in the city of Tequila in the state of Jalisco around the end of the 18th century. The Spaniards tried to supress the consumption of tequila through the issue of a decree signed by Carlos III forbidding the sale and production of the drink, as he considered it damaging for the native's health, although the real reason was that Tequila was competition for brandy and other wines imported from Spain. Since natives continued producing and drinking Tequila, the Spaniard governor of the region decided to authorize the production and collect royal tax. Around the end of the nineteenth century, the expansion of the Tequila industry was evident, but it was not until the first casks were exported to the U.S. that Tequila was known beyond Mexico's borders (6).

With the growing consumption of Tequila worldwide, several countries began selling other spirits as Tequila. Thus Mexican producers persuaded the government to stipulate that only the drink made with *Agave tequilana* weber, blue variety, according to certain quality standards could be called Tequila. By law, it is produced exclusively in certain areas of Mexico, including all of Jalisco and part of other states. Following Tequila's appellation of origin, the other distilled beverages, Mezcal, Bacanora, and Sotol, from the states of Oaxaca, Sonora, and Chihuahua, and other states, were granted equal recognition.

Although everyone prizes these complexely flavored spirits distilled from the juices of different agave plants, they have been scarcely studied. Some reports followed the development of flavor compounds during different steps in Tequila production, while others contributed to elucidate the complexity of the flavor profile in the final product. One report focused on the Maillard compounds generated under different conditions during the cooking step of *A. tequilana* weber var. azul for tequila production (7). Also, the use of different yeast strains during the fermentation step in tequila production were studied (8). Finally, the effect of the distillation step on the Tequila was reported (9). To gain knowledge of the chemical composition of Tequila flavor, dichloromethane extracts were analyzed by GC-MS and odor port evaluation/olfactometry (10-11). More than 175 components were identified in the extract, from which 60 odorants were detected. At least 30 compounds were correlated with specific GC peaks, however efforts at reconstituting Tequila flavor from its component parts were not successful, indicating that further significant contributors to Tequila flavor remain to be identified (10). In an effort to use a solvent-free sampling method, where sample handling is minimized, a solid phase microextraction sampling method (SPME) followed by GC/MS analysis of different tequila types was proposed (12).

This paper will address differences among Tequila, Mezcal, Bacanora, and Sotol with special emphasis on volatile characterization and aroma discrimination based on an electronic nose. To the best of our knowledge, this is the first paper that presents the chemical characterization of other Mexican distilled spirits (Mezcal, Bacanora, and Sotol) and discriminates the aroma profile of Tequila, Mezcal, and Bacanora, based on an electronic nose.

Chemical Characterization

Analysis of Volatile Constituents

Mexican regulations established certain limits for the presence of ethanol, methanol, higher alcohols, aldehydes, esters, and furfural in these agave spirits *(2-5)*. According to regulations, ethanol concentration (%) is allowed to vary in wide ranges, 38-55 for Tequila *(2)*, 36-55 for Mezcal *(3),* and 35-55 for Sotol. However, commercial spirits presented ethanol contents that ranged between 38-40%, except for Bacanora, which ranged between 42 and 46%. According to standards *(13),* the quantitative determination of methanol, higher alcohols, and aldehydes in alcoholic drinks, should be carried out by a gas chromatographic method using direct injection and a DB-WAX column (60 m x 0.25 mm i.d., 0.25 μm film thickness). Figure 1 shows a typical chromatogram of a mixture of analytical standards representing the analytes to be quantified in the spirits.

All volatiles were found to comply with regulations in Tequila (n = 29), Mezcal (n = 10), Bacanora (n = 8), and Sotol (n = 3) *(2-5, 13)*. These volatiles were present in the highest concentrations of all, after ethanol. Regulations also have standards for minor volatiles, such as esters and furfural. In fact, differences among types of spirits (white, gold, aged, and extra-aged) in the regulations are given based on total esters as determined by a volumetric method *(2-5, 13)*. However, important ester compositional differences cannot be obtained by this methodology *(12).* Thus a SPME-GC-MS method was reported that allowed Tequila volatile characterization and ethyl esters quantitation. A PDMS fiber (liquid inmersion) for volatile adsorption and a DB-5 column for the chromatographic runs were found suitable for the analysis *(12).*

Major volatiles identified in the different Tequila types were alcohols, esters, and ketones (Fig. 2). The most numerous volatiles detected were esters, and the most abundant volatiles after the alcohols were ethyl esters that showed qualitative and quantitative differences among Tequila types. Extra-aged Tequila presented the highest concentration of ethyl esters of all Tequila types (Table I). These results were expected since ethyl esters are formed not only during fermentation but also during aging. According to the authors, the SPME-GC methodology presented may be a good alternative for the classification of Tequila types. However, the analysis of multiple samples were required to be able to establish concentration ranges that allow correct classification *(12).* Similarly, the same ethyl esters ($C_6 - C_{18}$) were found in Mezcal, Bacanora, and Sotol as the most abundant volatiles after alcohols and the most numerous (Fig. 3).

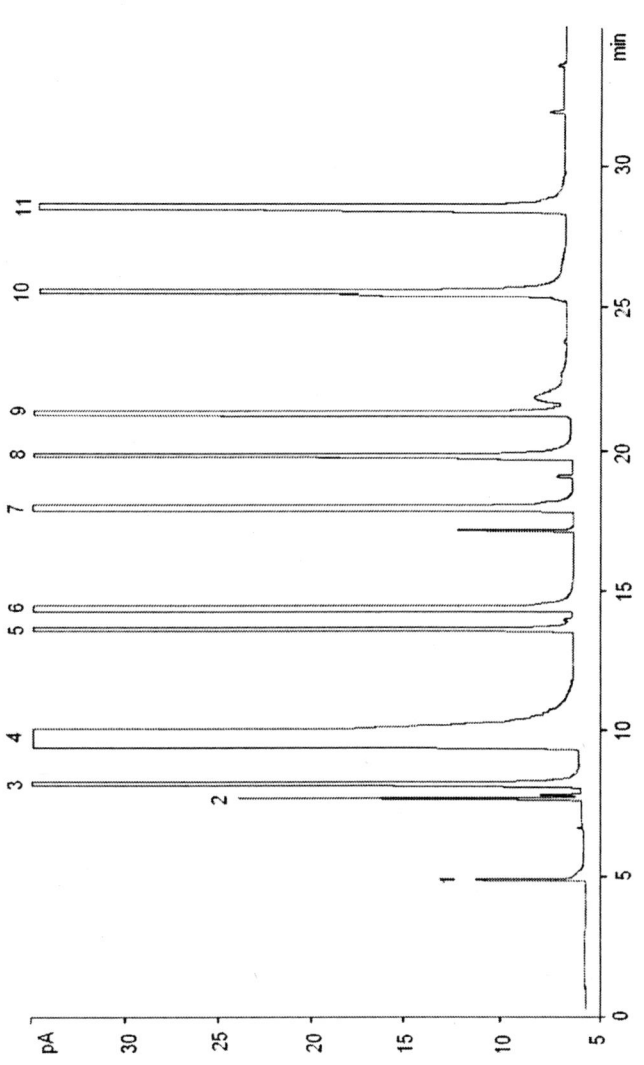

Figure 1. Typical chromatogram of volatiles determined by direct injection. Mixture of analytical standards:
(1) acetaldehyde; (2) ethyl acetate; (3) methanol; (4) ethanol; (5) 2-butanol; (6) 1-propanol; (7) 2-methyl-1-propanol;
(8) 2-pentanol (internal standard); (9) n-butanol; (10) 2-methy-1-butanol; (11) 1-pentanol.

158

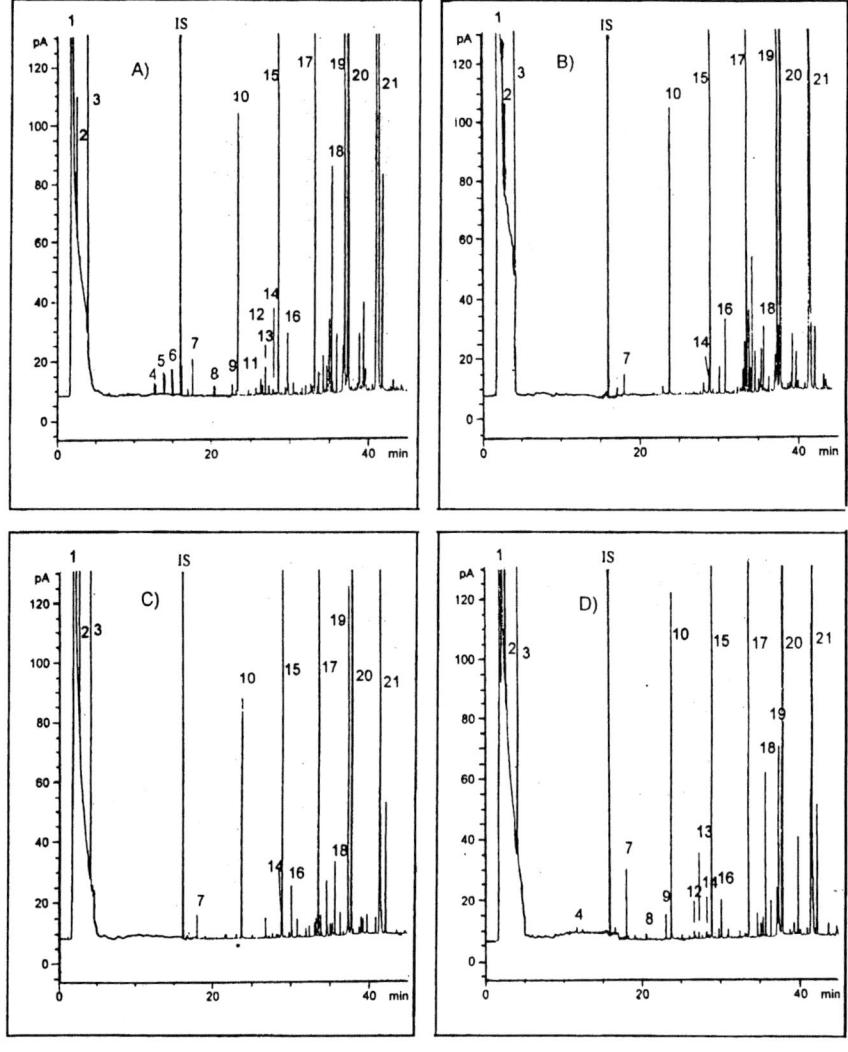

*Figure 2. Typical volatile profiles of (A) white, (B) gold, (C) aged, and (D)
extra-aged Tequila. Peaks: 1, ethanol[a]; 2, 1-propanol[a]; 3, 3-methylbutanol[b]; 4,
ethyl hexanoate[a]; 5, 4-methylheptanol[b]; 6, methyl heptanoate[a]; IS, internal
standard, methyl octanoate; 7, ethyl octanoate[a]; 8, methyl decanoate[a]; 9, 2-
buten-1-one[b]; 10, ethyl decanoate[a]; 11, 3-methyl butyl octanoate[b]; 12, Propyl
decanoate[b]; 13, methyl dodecanoate[a]; 14, butyl decanoate[b]; 15, ethyl
dodecanoate[a]; 16, 3-methyl ethyl decanoate[b]; 17, ethyl tetradecanoate[a]; 18, 2-
phenyl ethyl octanoate[b]; 19, 3-hexanone[b]; 20, ethyl hexadecanoate[a]; 21, ethyl
octadecanoate[a]. [a]Positively identified by SPME GC-MS and retention times of
authentic analytical standards. [b]Tentatively identified by SPME GC-MS.
Reproduced from reference 12. Copyright 2004 Amercian Chemical Society.*

Furfural or 2-furaldehyde in agave spirits is a degradation product of the hydrolysis of pentoses during the cooking of agave, and also may leach from oak barrels during storage and aging. The determination of furfural in these spirits represents a major challenge since the traditional spectrophotometric method established in the regulations *(13)* has the disadvantage of instability of the colored complex formed and the time required for analysis.

Thus an alternative capillary electrophoresis technique that was accurate, reproducible, and fast was developed (unpublished). A micellar electrokinetic capillary electrophoresis (MEEKC) method using solid phase extraction (SPE) sample pretreatment and detection at 280 nm was used. A typical electropherogram showing the furfural peak in Bacanora (Fig. 4A) and Bacanora spiked with the analytical standard (Fig. 4B) are shown. Commercial samples of Tequila or Mezcal analyzed by the newly developed method presented furfural concentrations of less than 1 ppm, whereas Bacanora presented concentrations slightly higher (1.2-1.3 ppm), with all samples found to comply with regulations *(2-4, 13)*.

Electronic Nose

An electronic nose, which is a sensor array-based instrument that emulates the human nose, was used to discriminate among Tequila, Mezcal, and Bacanora. A commercially available electronic nose consisting of 12 gas sensors and a pattern recognition algorithm was used (Fox 3000, Alpha Mos). The headspace generated by samples heated at 35°C was transfered to the detection device which revealed differences in sensor reactions using discriminant factorial analysis. Good discrimination between different Tequila types was shown (Fig. 5). According to regulations, the product known as white Tequila is clear with no aging, produced from a fermented wort containing not less than 51% sugars from the agave plant *(2)*. Aged and extra-aged Tequila are white Tequila matured in wood containers or oak casks for at least two and twelve months, respectively *(2)*. Ancient Tequila, which is different from the rest, was a specific type of Tequila manufactured by a distillery. Close proximity among Tequila and Mezcal samples, with Bacanora being distinctively different was observed when all spirits were analyzed (Fig. 6).

Bacanora is still 100% artisanal, starting from the cooking step which is still done in holes underground. Fermentation is carried out using wild yeasts and distillation is accomplished using rustic alembics, while Tequila and Mezcal are industrialized. Volatiles and hence the aroma produced in the final products are

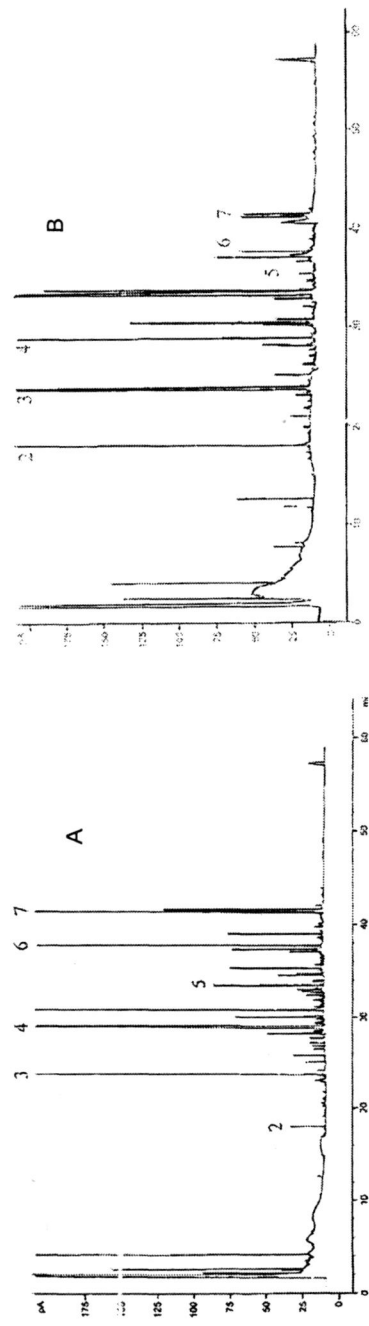

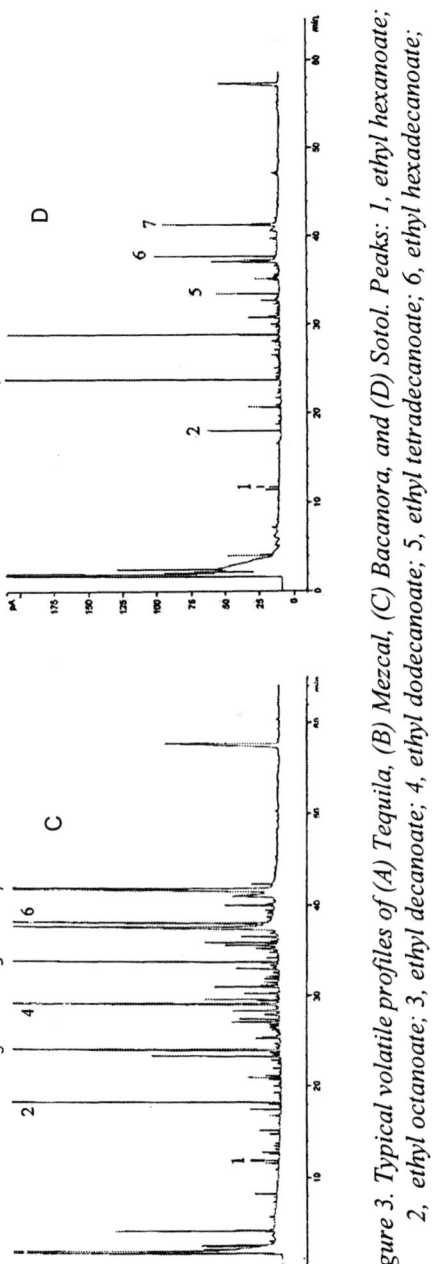

Figure 3. Typical volatile profiles of (A) Tequila, (B) Mezcal, (C) Bacanora, and (D) Sotol. Peaks: 1, ethyl hexanoate; 2, ethyl octanoate; 3, ethyl decanoate; 4, ethyl dodecanoate; 5, ethyl tetradecanoate; 6, ethyl hexadecanoate; 7, ethyl octadecanoate.

Table I. Quantitative Determination of Ethyl Esters Present in Tequila

Ethyl Esters	Gold[a]		White[a]		Aged[a]		Extra-Aged[a]	
	Mean (ppm)	CV[b] (%)	Mean (ppm)	CV[b] (%)	Mean (ppm)	CV[b] (%)	Mean (ppm)	CV[b] (%)
Hexanoate	0.27	-	ND	-	ND	-	ND	2.0
Octanoate	0.65	1.4	0.62	1.2	0.70	1.0	1.98	1.6
Decanoate	3.54	2.0	4.25	1.5	4.00	1.5	4.33	0.90
Dodecanoate	3.25	2.3	3.18	2.2	3.73	2.4	5.97	1.7
Tetradecanoate	1.09	5.3	0.73	4.8	0.87	5.3	2.61	4.5
Hexadecanoate	7.42	6.4	8.73	6.8	9.95	7.0	13.08	7.6
Octadecanoate	9.85	8.1	10.36	8.7	11.90	8.5	15.03	9.8

[a]Extraction conditions for ethyl esters, tequila (40 mL) at 40 °C; sampling mode; direct liquid; fiber PDMS; sampling time, 60 min. [b]CV, coefficient of variation, n = 3. [c]ND = not detectable. Reproduced from Reference 12. Copyright 2004 American Chemical Society.

mainly the result of cooking, fermentation and distillation steps during production of these spirits. Thus, it is reasonable to observe similarities between Tequila and Mezcal which are industrialized, while Bacanora, which is artisanal, is very different (Fig. 6).

Conclusions

Although the flavor of Tequila has conquered international palates and shares the popularity of cognac, whiskey, or brandy, its complex, delicate, and exquisite aroma and taste have been scarcely studied. Aroma notes such as floral, woody, fruity, herbal, smoky, and sweet were detected by sniffing GC ports, differences among Tequila types were detected by an electronic nose, and major and minor volatiles in Tequila were characterized by GC and capillary electrophoresis. However, there is need for further chemical and sensory characterization of Tequila during production steps, as well as in the final product. Although Mezcal, Bacanora, and Sotol are not as popular as Tequila, their study is equally important to determine differences based on raw materials and processes.

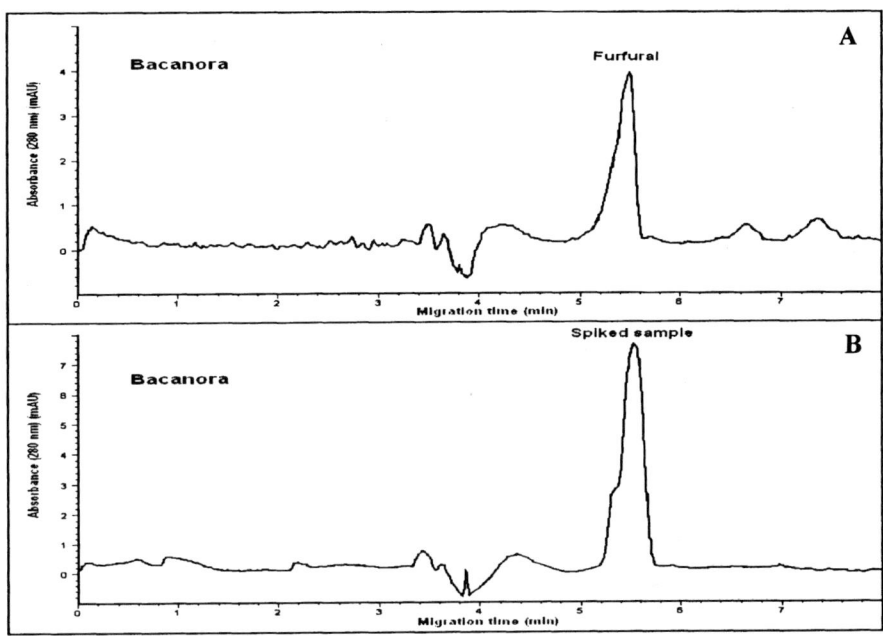

Figure 4. Typical electropherograms of (A) Bacanora and (B) Bacanora spiked with the analytical standard.

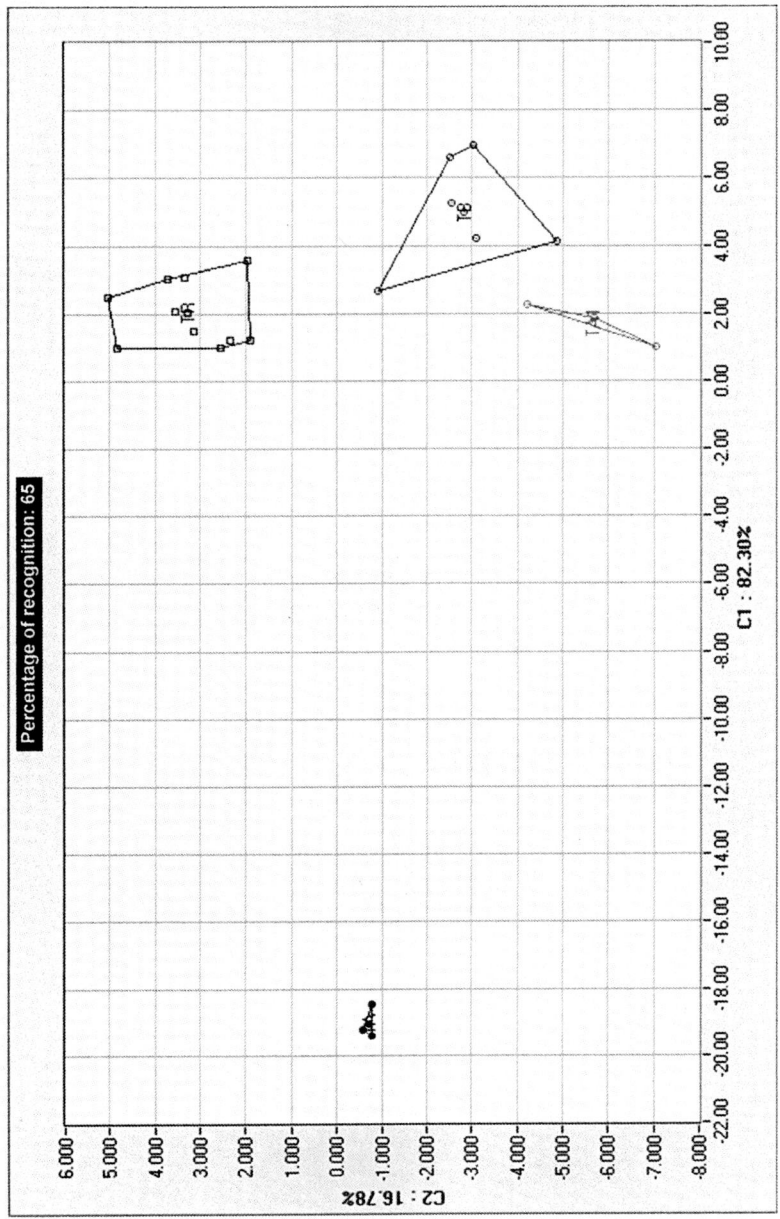

Figure 5. Differentiation of Tequila types based on electronic nose.

165

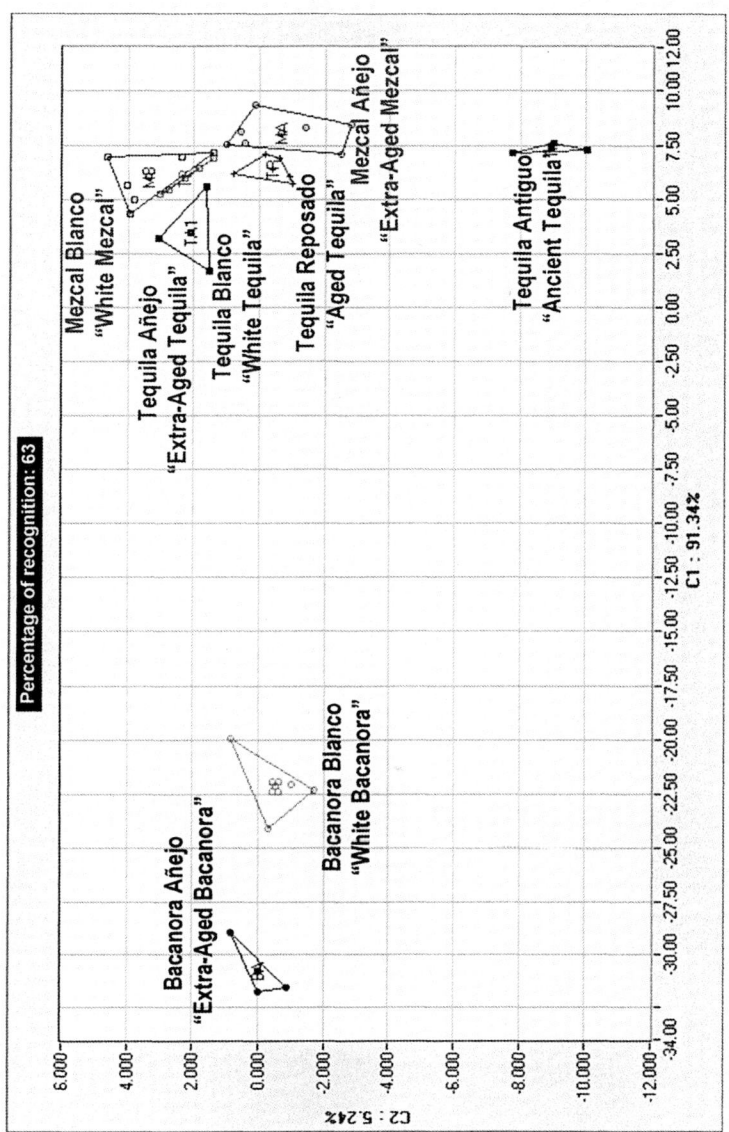

Figure 6. Differentiation of Tequila, Mezcal, and Bacanora based on electronic nose.

166

Acknowledgments

The authors wish to thank Dr. Aideé Orozco from Tequila Herradura, S.A. de C.V. (Amatitán, Jalisco) and Ing. Alejandro Luján from MACROLAB Instrumentos, S.A. de C.V. (Monterrey, Nuevo León). The technical assistance of María del Carmen Estrada-Montoya and Roberto Rodríguez-Ramírez was appreciated.

References

1. Iñiguez-Covarrubias, G.; Lange, S.E.; Rowell, R.M. *Bioresource Technol.* **2001**, *77*, 25-32.
2. Mexican Ministry of Commerce and Industry. Regulations: NOM-006-SCFI-1994. Alcoholic Drinks -Tequila- Specifications; Diario Oficial de la Federación: México, Septiembre 3, 1997.
3. Mexican Ministry of Commerce and Industry. Regulations: NOM-070-SCFI-1994. Alcoholic Drinks -Mezcal- Specifications; Diario Oficial de la Federación: México, Agosto 17, 1994.
4. Mexican Ministry of Commerce and Industry. Regulations: PROY-NOM-168-SCFI-2004. Alcoholic Drinks -Bacanora- Specifications; Diario Oficial de la Federación: México, Agosto 17, 1994.
5. Mexican Ministry of Commerce and Industry. Regulations: NOM-159-SCFI-2004. Alcoholic Drinks -Sotol- Specifications; Diario Oficial de la Federación: México, Diciembre 6, 2002.
6. Cedeño, M. *Crit. Rev. Biotechnol.* **1995**, *15*, 1-11.
7. Mancilla-Margalli, N.A.; López, M.G. *J. Agric. Food Chem.* **2002**, *50*, 806-812.
8. Pinal, L.; Cedeño, M.; Gutiérrez, H.; Alvarez-Jacobs, J. *Biotechnol. Lett.* **1997**, *19*, 45-47.
9. Prado-Ramírez, R.; Gonzáles-Alvarez, V.; Pelayo-Ortiz, C.; Casillas, N.; Estarrón, M.; Gómez-Hernández, H.E. *Int. J. Food Sci. Tech.* **2005**, *40*, 701-708.
10. Benn, S.M.; Peppard, T.L. *J. Agric. Food Chem.* **1996**, *44*, 557-566.
11. López, M.G.; Dufour, J.P. Tequilas: Charm analysis of blanco, reposado and añejo tequilas. In *Chromatography-Olfactometry. The State of the Art*; Leland, J.V.; Schieberle, P.; Buettner, A.; Acree, T.E.; Eds.; ACS Symposium Series 782; American Chemical Society: Washington, DC, 2001; pp 62-72.
12. Vallejo-Cordoba, B.; González-Córdova, A.F.; Estrada-Montoya, M.C. *J. Agric. Food Chem.* **2004**, *52*, 5567-5571.
13. Mexican Ministry of Health. Regulations: NOM-142-SSA1-1995. Alcoholic Drinks Specifications; Diario Oficial de la Federación: México, Junio 3, 1996.

Chapter 14

Production of Volatile Compounds in Tequila and Raicilla Musts by Different Yeasts Isolated from Mexican Agave Beverages

Javier Arrizón, José Javier Arizaga, Rosa Elia Hernandez, Mirna Estarrón, and Anne Gschaedler

Centro de Investigación y Asistencia en Tecnologia y Diseño del Estado de Jalisco, A.C., Guadalajara, Jalisco, Mexico 44270

In order to find yeasts capable of producing terpenoids and volatile compounds during raicilla and tequila fermentation, a screening for the presence of the enzymes β-glucosidase, β-xylosidase, and β-cellobiosidase was performed from yeasts isolated from mezcal (San Luis Potosí type), pulque, and raicilla. In addition, an evaluation of the production of some of the volatile compounds important in distilled beverages was performed on *Agave tequilana* must (tequila) and *Agave maximiliana* must (raicilla), with 10 microorganisms selected. Finally, a comparison of the terpenoids in raicilla and tequila final products was performed. From 218 microorganisms isolated, only 20 were positive to the three enzymes (11 from pulque, 8 from raicilla, and 1 from mezcal). The majority of the yeasts isolated from raicilla did not contain β-xylosidase, whereas the ones isolated from mezcal and pulque contained the three enzymes. When a comparison of the terpenoids identified in tequila and raicilla final distilled products was carried out, 16 terpenoids were present in both beverages and 15 were only present in raicilla.

Different agave distilled beverages such as tequila, mezcal, sotol, bacanora, and raicilla are produced in Mexico. The production process is similar with all of these beverages, but there are differences in practices, principally between tequila and the other distilled beverages. In the case of tequila, only the *Agave tequilana* blue variety is used; the production is at an industrial level with more technology, and the fermentation can occur with or without inoculation. In the case of mezcal, sotol, bacanora, and raicilla, a great variety of agave plants are used as raw material, and artisan procedures are normally used with less control during the process, where natural fermentation occurs. These differences in procedures and raw material have an impact on the production of volatile compounds, which is related to the sensory quality of the beverage. In this work, the production of volatile compounds during fermentation of tequila and raicilla was compared.

Mexican Distilled Beverages Obtained from Agave

Principal Agave Beverages Produced in Mexico

In general, there are two kinds of agave beverages, non-distilled and distilled. Pulque is the only non-distilled beverage, and is produced by the fermentation of the juice of the *Agave salmiana* plant, without cooking. Agave distilled beverages can be divided into beverages produced by industrial processes such as tequila, and artisan processes like mezcal, sotol, bacanora, and raicilla. The generic name for all these distilled agave beverages is "mescal," which means the product of distilling the fermented must of different agaves. The agave species used are different according to each region (Table I). The process to obtain a mezcal type beverage involves several steps. First, the leaves are taken off, resulting in clean agave heads. Agave heads are then cooked to hydrolyze the polysaccharides (mainly inulin) into fermentable sugars, a step which is performed in stone or brick ovens for artisan processes and in an autoclave or mamposteria oven for industrial processes. Cooked heads are milled either by hand or with tahona (a stone device) for artisan processes, and by milling devices in the industrial processes. Once the juice is extracted, fermentation occurs. For artisan processes this is always carried out by natural fermentation without yeast inoculation, and for industrial processes fermentation can be performed with and without yeast inoculation. In the case of tequila, the origin of the yeasts can be from wine or baking, or yeast strains could be isolated from agave fermented musts. After fermentation, distillation is carried out with rustic devices for artisan processes and with a pot still or distillation column for the industrial processes. In some cases a process of aging is carried out after distillation.

Table I. Regions of Production and Type of Agave Used for the Different Distilled Beverages in Mexico

Agave beverage	Raw material	Region
Tequila	*A. tequilana* blue variety	Jalisco State and some municipalities in west and northeast of Mexico
Mezcal	*A. durenguensis*	Durango State
Mezcal	*A. salmiana*	San Luis Potosí State
Mezcal	*A. salmiana* and *A. tequilana*	Zacatecas State
Mezcal	*A. potatorum* and *A. angustifolia*	Oaxaca State
Mezcal	*A. cupreata*	Guerrero State
Sotol	*Dacilirium*	Chihuahua State
Bacanora	*A. angustifolia*	Sonora State
Racilla	*A. angustifolia, A. maximiliano* and *A. inaequidens*	Jalisco State
Pulque	*A. salmiana*	States in the center of Mexico

Raicilla and Tequila Beverages from Jalisco

In the 17th century, Fray Toribio de Benavente described for the first time the beverage called "mexalli." He wrote, "Spanish said that this beverage is strong and healthy" (*1*). But the real father of the mezcal beverage is the Spaniard Pedro Rodrígeuz de Alberme, who introduced the distillation process in New Spain to the indigenous agave fermented beverages of Mexico (*2*).

It is important to point out that in ancient times, the production of distilled beverages from agave was forbidden by the Virreinal authority. In 1785 the sale of any alcoholic beverage was forbidden in New Galicia, but finally in 1795 Don José María Guadalupe Cuervo was allowed to produce the distilled beverage called Tequila (*2*). Tequila is produced principally in Jalisco State and some municipalities of Guanajuato, Nayarit, Michoacan, and Tamaulipas, and these production regions are protected by an agreement of origin.

There are two kind of raicilla, "raicilla de la sierra" and raicilla de la costa," the principal difference between them being the agave plants used. Raicilla is only produced in the western part of Jalisco State.

Properties of Yeasts Isolated from Fermentation

A great diversity of yeasts participate at the beginning of the fermentation process, after which a few microorganisms predominate, such as *Saccharomyces cerevisiae* in tequila (*3*) and *S. cerevisiae* and *Candida spp.* in Mezcal (*4*). This complex interaction between the microorganisms makes the production of a great number of volatile compounds possible. In order to facilitate the study of the fermentation properties of these yeasts, they are divided in two groups: *Saccharomyces* and non-*Saccharomyces* yeasts.

Non-*Saccharomyces* yeasts isolated from different agave musts are suitable for tequila fermentation. They can also be used in the wine industry due to their high ethanol and sulfite tolerance, which is not common for non-*Saccharomyces* yeasts isolated from wine must (*5*). It has been also observed that a non-*Saccharomyces* strain of *Candida magnoliae* isolated from mezcal produces as much ethanol as *Saccharomyces* strains. The *Saccharomyces* strains isolated from agave must are also capable of fermenting wine, such as wine *Saccharomyces* strains (*6*), which make these strains interesting for the alcoholic beverage industry.

In tequila fermentation, it has been found that the yeast strain (*6-8*), the must composition (*8*), the method of supplementing the nitrogen source (*9*), and the carbon/nitrogen ratio (*7*) are important for the production of volatile compounds. In experiments where the *Agave tequilana* musts were supplemented at the beginning of the fermentation with organic and inorganic sources of nitrogen using a high alcohol producer (*Saccharomyces cerevisiae*) or a high aroma producer (*Kloeckera africana*), the fermentation capacity increased in the *Saccharomyces* strain using the inorganic source, whereas in the *Kloeckera* strain the fermentation performance increased with organic ones. The aroma profile also varied depending on the strain and the nitrogen source used (*7*).

In fermentations with *Saccharomyces cerevisiae*, at high *Agave tequilana* sugar concentration (170 g/L) and with the nitrogen source (70% inorganic, 30% organic) added in the exponential growth phase, the fermentation rate and efficiency increased. The biosynthesis of higher alcohols, namely isobutanol and amylic alcohols, decreased whereas the biosynthesis of propanol increased (*9*).

The production of volatile compounds in tequila fermentation also depends on the concentration of sugar used. Tequila fermentation in the industry is performed at 12° Brix (100-120 g/L of sugar) with yeasts from different origins (wine, bakery, rum, etc.) and in a few cases with yeasts isolated from agave musts. In order to produce more tequila in the same time, the industry has to perform fermentation at high sugar concentration. A comparison of volatile compound production by *Saccharomyces*/non-*Saccharomyces* strains from wine and from agave musts (tequila, mezcal, sotol, and raicilla) was performed in *Agave tequilana* must at a very high sugar concentration (300 g/L). It was observed that wine strains were strongly inhibited, and as a result only a few volatile compounds were produced, and in particular the higher alcohol 2-phenylethanol was not produced with all the wine strains tested (*6*). This compound is very important in the sensory quality of tequila (*10*). Isobutanol and amylic alcohols were present at a higher level in tequila when the C/N ratio was low (62/1), compared with a high C/N ratio (188/1) (*7*).

In most of the beverages, there are compounds like terpenes at very low concentrations that give the distinctive aroma characteristics. The terpenes can be present in the vegetal material used to make the must, or they can be delivered by the action of the microbial enzymes like β-glucosidase, β-cellobiosidase, and β-xylosidase, which break the chemical bonds between terpenes and the polysaccharides such as cellulose and hemicellulose. A screening for the presence of β-glucosidase and β-xylosidase has been performed on 6 yeasts isolated from agave musts. It was found that *Saccharomyces* strains contain β-xylosidase and the non-*Saccharomyces* strains contain both enzymes, but β-glucosidase has more activity in one yeast of *Candida magnoliae* (*5*).

Production of Volatiles from Yeasts

In the agave distilled beverages as in other alcoholic beverages, esters, higher alcohols, and some compounds present in very low quantities such as terpenes are important for the bouquet. In the particular case of tequila, Benn and Peppard (*10*) found that amylic alcohols and 2-phenylethanol are important for the characteristic tequila aroma. Nevertheless there is little information about the role of terpenes in the aroma of agave distilled beverages. In order to find yeasts with the potential of terpene production in tequila and raicilla fermentations, a screening for enzymes involved in terpene production (β-glucosidase, β-cellobiosidase, and β-xylosidase) was carried out. Yeasts were selected from different factories: one pulque factory, one mezcal factory from San Luis Potosí State, and six different raicilla factories. From 218 microorganisms isolated, 20 were positive for the three enzymes (Table II).

Table II. Results of β-Glucosidase, β-Cellobiosidase, and β-Xylosidase Screening of Yeasts Isolated from Different Agave Beverages

Source	Yeasts Isolated	Yeasts Testing Positive for at Least One of the Three Enzymes	
		Number	Percentage
Raicilla	182	8	4.3
Pulque	33	11	33.3
Mezcal	3	1	33.3

Figure 1 shows that the strains isolated from raicilla must did not contain the enzyme β-xylosidase, whereas the strains isolated from pulque and mezcal contained all three enzymes. "Raicilla de la Sierra" is made from *Agave maximiliana,* and pulque and mezcal are made from *Agave salmiana.* The reason for the differences found in the presence of the enzymes from the different yeasts could be that *Agave maximiliana* did not contain terpenes bounded to xylan. Therefore the microorganisms which adapted to this fermentation medium did not contain β-xylosidase, whereas the *Agave salmiana* contained such terpenes bound to both cellulose and xylan. Hence, the yeasts in mescal and pulque contained the enzymes.

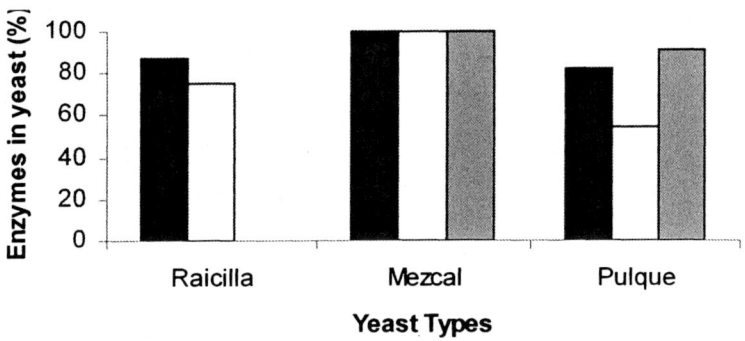

Figure 1. Percentages of the enzymes β-glucosidase (black bars), β-cellobiosidase (white bars), and β-xylosidase (gray bars) in the yeast strains isolated from raicilla, mescal, and pulque.

The capacity to produce acetaldehyde, ethyl acetate, and higher alcohols (2-phenylacetate, 1-propanol, 2-methylpropanol, and amyl alcohols) was evaluated with ten of the twenty yeasts containing the mentioned enzymes. Figures 2 and 3 show that the strains isolated from pulque did not ferment any of the musts tested, therefore volatiles were not produced. The strains of raicilla produced more volatile compounds in raicilla must than in tequila must. From the strains of mezcal, the strain 216 fermented tequila must but not raicilla must (Fig. 3), and the strain 217 did not ferment any of the musts (Figs. 2 and 3).

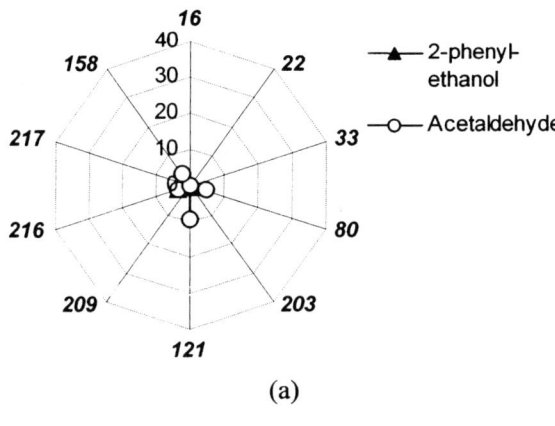

(a)

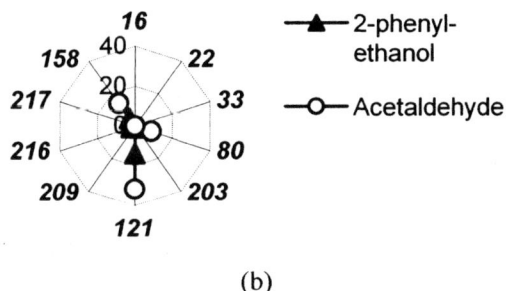

(b)

Figure 2. Minor volatile compound production (mg/L) in (a) tequila and (b) raicilla fermented musts with strains isolated from mezcal (yeasts 216 and 217), pulque (yeasts 16, 22, and 33) and raicilla (yeasts 80, 203, 121, 209, and 158).

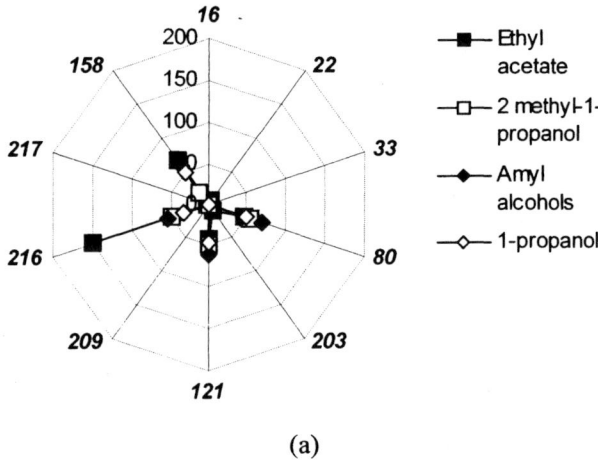

(a)

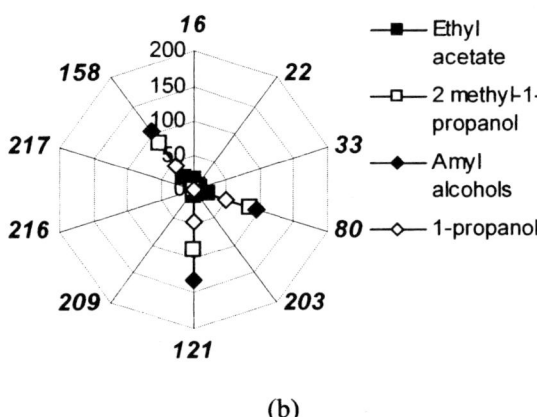

(b)

Figure 3. Major volatile compound production (mg/L) in (a) tequila and (b) raicilla fermented musts with strains isolated from mezcal (yeasts 216 and 217), pulque (yeasts 16, 22, and 33) and raicilla (yeasts 80, 203, 121, 209, and 158).

Therefore, it can be assumed that the raicilla must has selectivity over the yeasts isolated from other agave musts. The same effect has been observed with strains isolated from grape wine and from agave musts in *Agave tequilana* juice fermentation (*6*).

Comparison of Volatiles in Tequila and Raicilla Final Products

Table III compares the presence of volatile terpenes in the final distilled tequila and raicilla products.   Chromatographic analysis of terpenes after liquid-

Table III. Terpenoids in Raicilla and Tequila Extracts

Terpenoids	Raicilla "Mascota"	Raicilla "Chaleco"	White Unaged Tequila
α-Pinene	++	n.d.	
Camphene	++	n.d.	
α-Terpinene	++	+	
β-Myrcene	+	++	x
Isocineole	trace	++	x
Limonene	++	+	
γ-Terpinene	++	+	
cis-β-Ocimene	++	n.d.	
p-Cymene	++	+	
α-Terpinolene	++	n.d.	
trans-β-Ocimene	++	n.d.	
cis-Linalool oxide	++	+	x
trans-Linalool oxide	++	+	x
Cycloisosativene	++	n.d.	
Linalool	++	+	x
1-Terpineol	++	++	x
4-Terpineol	++	+	x
Mycenol	+	++	
β-Terpineol	++	+	x
α-Terpineol	++	+	x
δ-Cadinene	++	trace	x
Neryl acetate	+	++	
Geranyl acetate	++	++	
Citronellol	++	+	x
Nerol	+	++	x
Geraniol	++	+	x
Nerolidol	++	+	x
δ-Selinene	++	++	x
α-Eudesmol	++	+	
Farnesol	++	+	x
β-Eudesmol	+	+	

++ Major amount; + minor amount; n.d.: not detected; x: compound commonly found in white tequila (11).

liquid extraction was performed in two different raicilla products, "Raicilla de la Sierra" from the Mascota region, and from the Chaleco region. The terpene composition was compared with the terpenes that are normally found in white tequila without aging (*11*). The two distilled products of raicilla must had similar terpene composition, even though they were fermented in different factories. As seen in Table III, some terpenes are common in tequila and raicilla and others could only be found in raicilla. Thus analysis of terpenes in distilled products could be useful for determining the authenticity of these beverages.

Conclusions

In this work, the differences in fermentation and volatile compound production between yeasts depended on the agave must tested. The yeasts isolated from raicilla must did not show the presence of β-xylosidase; therefore there are differences in the composition of terpene precursors between the different agave musts used. As a result, there are different terpene concentrations in the tequila and raicilla final products. The differences in composition between the different agaves used probably affects the yeasts, and thus by an adaptation process some enzymes are not present in the yeasts. In the future, a study has to be performed on the generation of volatiles in the distilled beverages from agave during the complete production process, from the agave plant until the distillation step, in order to understand which factor is the most important in the generation of terpenes and other volatile compounds that are key to the sensory quality of the beverage. Finally, an evaluation of other agave musts on the fermentation behavior of yeasts isolated from other sources has to be performed in order to test the selectivity of the agave musts on the yeasts.

References

1. Muria, J.M.; Sánchez, R. In *Una Bebida Llamada Tequila*; Editorial Agata: Mexico City, Mexico, 1997; pp 1-25.
2. Muria, J.M. In *El Tequila, Arte Tradicional de México*; Las Artes de México: Mexico City, Mexico, 1994; pp 28-36.
3. Lachance, M.A. *Anton. Leeuw.* **1995**, *68*, 151-160.
4. Andrade-Meneses, O.E.; Ruiz-Terán, F. *Eleventh International Congress on Yeast* **2004**, Rio de Janeiro, Brazil.
5. Fiore, C. ; Arrizon, J. ; Gschaedler, A. ; Flores, J.; Romano, P. *W. J. Microbiol. Biotechnol.* **2005**, in press.

6. Arrizon, J.; Acosta, G.; Gschaedler, A.; Fiore, C.; Romano, P. *Eleventh International Congress on Yeast* **2004,** Rio de Janeiro, Brazil.

7. Pinal, L.; Cedeño, M.; Gutiérrez, H.; Alvarez-Jacobs, J. *Biotechnol. Lett.* **1997,** *19,* 45-47.

8. Diaz-Montaño, D.; Delia, M.L.; Strehaiano, P.; Gschaedler, A., 22nd *Internacional Specialised Symposium on Yeast* 2002, Pilanesberg National Park, South Africa.

9. Arrizon, J.; Gschaedler, A. *Can. J. Microbiol.* **2002,** *48,* 965-970.

10. Benn, S.M.; Peppard, T.L. *J. Agric. Food Chem.* **1996,** *44,* 557-566.

11. Escalona-Buendía, H.B.; Villanueva-Rodríguez, S.J.; López, J.E.; Gonzáles, R.M.C.; Martín del Campo, T.; Estarrón, M.; Cosio, R.; Cantor, E. In *Ciencia y Tecnología del Tequila; Avances y Perspectivas;* CIATEJ: Guadalajara, Mexico, 1994; pp 173-256.

Chapter 15

The Flavor of the Classic Margarita Cocktail

Sanja Eri, Neil C. Da Costa, and Laurence Trinnaman

International Flavors and Fragrances Inc., 1515 Highway 36,
Union Beach, NJ 07735

The classic Margarita cocktail is made with tequila, orange liqueur, and lime juice. Among other things, this cocktail owes its popularity to its smooth, pleasant, and refreshing flavor. Each of the main ingredients makes its own contribution to the overall perception of this unique flavor. One way to investigate the flavor of a classic Margarita cocktail is to look at the flavor profiles of its ingredients. Tequila, orange, and lime were analyzed as part of the Generessence® program at International Flavors & Fragrances, Inc. Their flavor compositions were determined by gas chromatography-mass spectrometry. Three sample preparation methods were used in the analysis: liquid/liquid extraction, dynamic headspace, and stir bar sorptive extraction (Twister™). The flavor profiles were assessed by combining the results obtained by different sample preparation methods. Each of the main ingredients has been shown to contain few hundred flavor components, indicating that a classic Margarita cocktail flavor is a very complex blend.

There are several stories describing how the classic Margarita cocktail was invented and how it got its name. All of them date back to 1940's and are placed in Mexico, hence the place and time seems to be consistent. Whether it was first consumed in Rancho La Gloria bar in Tijuana and named after an actress who used to go there, or in the Acapulco villa of socialite Margarita Sames still needs to be resolved. Or maybe not, since it adds to the attractiveness of the cocktail and makes a nice conversation piece while enjoying it. In any case, this cocktail has become known and enjoyed throughout the world. One of the reasons for its widespread popularity is its smooth and refreshing flavor. The Classic Margarita Cocktail recipe calls for 1 oz of tequila, 1 oz of lime juice, and ½ oz of orange liqueur. Ingredients are usually combined in a cocktail shaker with ice which can be in cubes, coarsely chopped, or finely crushed. The mixture can be strained or poured into a Margarita or Martini glass which may be rimmed with salt and garnished with a wedge of lime. The end result is a pleasant flavor to which each of the ingredients contributes in its own way. Tequila is an alcoholic beverage produced from the juice of blue agave plant grown in Mexico. Based on the length of aging period, there are three types of tequila: Blanco (white), which is aged less than a month, Reposado (rested), which is aged minimum two months but less than a year, and Anejo (aged) which is aged for a year or more. They differ in intensity of agave flavor, Blanco tequila being the strongest. Tequila aroma has been the subject of several investigations *(1, 2)*. The second ingredient which provides freshness to the Margarita flavor is lime juice *(3)*. Orange liqueur is a beverage made by infusion or maceration of orange peel *(4)* and spirit such as brandy, cognac, etc. Depending on the spirit and process used there are several types (Cointreau[TM], Grand Marnier[TM], Triple Sec[TM], Curacao[TM], etc.) of orange liqueur.

Since each of the ingredients in the classic Margarita makes its own contribution to the flavor of this cocktail, one way to study it would be to look at the flavors of tequila, lime juice, and orange liqueur.

As part of the Generessence® program at International Flavors & Fragrances Inc., we have analyzed Blanco tequila, Persian limes, and Valencia orange peel oil. The program involves the in-depth analysis of a natural product to enable the creation of closer-to-nature flavors, using only the chemicals found in the analyses. An additional aim is the discovery of novel compounds in the natural product and occasionally in nature so they may be synthesized and ultimately registered for use. Therefore, as much detail as possible is required for the analysis of volatiles, semi-volatiles, and even non-volatiles present. Many extraction and isolation techniques are available to the analyst and are well documented in the literature *(6, 7)*. In this particular case the extraction techniques used were liquid/liquid extraction *(8)*, stir bar sorptive extraction (Twister[TM]) *(9, 10)* and headspace sampling *(11, 12)*. Liquid/liquid extraction gives a good recovery of aroma compounds and it is the most quantitative of all flavor isolation methods. Stir bar sorptive extraction and headspace were chosen to complement the liquid/liquid extraction. Sampling by stir bar sorptive extraction is particularly useful for the analysis of alcoholic beverages since the alcohols which mask the early part of the chromatogram on a non-polar phase appear at much lower concentrations revealing trace components, if present. Headspace trapping of the aroma was conducted using Tenax™ traps to capture the highly volatile compounds.

Materials and Methods

Liquid/liquid extraction of tequila (Blanco tequila) and lime juice (freshly squeezed juice of Persian limes) was performed by using dichloromethane (Fischer Scientific Co., Springfield, NJ) as a solvent. Limes were clean from the rind and liquefied in an electric juicer which removed much of the pulp material. The juice was then extracted with dichloromethane. Both extracts were dried over anhydrous sodium sulfate, filtered, and reduced first by a rotary evaporator and then under nitrogen to a volume of 1 mL. The extracts were analyzed by a gas chromatograph fitted with a flame ionization detector (GC-FID), and gas chromatography-mass spectrometry (GC-MS).

For the stir bar sorptive extraction, teqilla and lime juice were put into Twister vessels (Gerstel Inc., Baltimore, MD) and on a stirrer plate. The Twister bars were added and volatiles were collected for 2 hr and 4 hr respectively, with spinning at 1800 rpm. The stir bars were rinsed with distilled water, padded dry with tissue paper and thermally desorbed onto GC-FID and GC-MS systems.

For the headspace analysis of lime juice, juice sample was prepared as for the liquid/liquid extraction and placed into a glass chamber. The sample was purged with helium at room temperature and volatile components were trapped onto Tenax TA traps (Supelco, Inc., Bellefonte, PA) using a vacuum pump (Fischer Scientific, Co., Springfield, NJ) with a flow rate of 25 mL/min. The headspace was sampled over 2-hr and 4-hr periods, respectively. The traps were thermally desorbed onto GC-FID and GC-MS systems.

The orange oil was obtained at an orange processing plant.

Gas Chromatography

The liquid extracts were analyzed using a HP6890 gas chromatograph with a split/splitless injector and a flame ionization detector (FID) (Hewlett Packard, Wilmington, PA). The extracts were injected onto a non-polar OV-1 capillary column (50 m x 0.32 mm i.d., 0.5 μm film thickness, Restek, Bellefonte, PA) in the split (split ratio 40:1) and splitless modes. The carrier gas was helium with a flow rate of 1.0 mL/min operated at constant flow. The injection port temperature was 250°C and the detector temperature was 320°C. The column temperature was ramped from 40°C to 270°C at 2°C/min with a 10 min hold at the final temperature.

A polar chromatogram of the extract was also obtained by injecting onto an HP5890 gas chromatograph with split/splitless injection and a flame ionization detector (FID) fitted with a Carbowax capillary column (50 m x 0.32 mm i.d., 0.3 μm film thickness, Restek, Bellefonte, PA) using the previously described injection and detection techniques. The temperature program began with an initial temperature of 60°C held for 10 min, ramped at 2°C/min to a final temperature of 220°C, and held for 20 min.

Tenax TA traps and Twister bars were thermally desorbed onto an HP6890 gas chromatograph equipped with a flame ionization detector (FID) using a Gerstel thermal desorber Model TDS 2 (Gerstel Inc., Baltimore, MD). Desorption time was 5 min at 250°C. The column was an OV-1 capillary column and the analysis was performed in splitless mode. The injector temperature was programmed from -150°C (held for 5 min during the thermal desorption) to 250°C. The detector temperature was 320°C.

All data were collected and stored by using HP ChemStation software (Hewlett Packard, Wilmington, PA).

Mass Spectrometry

Identification of components in the extracts was conducted by mass spectrometry. The sample was injected onto an HP5890 GC. The chromatographic conditions for the OV-1 column were the same as described for GC analysis. The end of the GC capillary column was inserted directly into the ion source of the mass spectrometer via a heated transfer line maintained at 280°C. The mass spectrometer was a Micromass Prospec high resolution, double-focusing, magnetic sector instrument. The mass spectrometer was operated in the electron ionization mode (EI), scanning from m/z 450 to m/z 33 at 0.3 s per decade.

Polar GC-MS analysis was on conducted on a Carbowax capillary column (50 m x 0.32 mm i.d., 0.3 μm film thickness); the sample was introduced via an HP5890 GC into a Kratos Profile mass spectrometer (Manchester, UK). GC oven conditions were the same as outlined above. The mass spectrometer was operated in EI mode scanning from m/z 450 to m/z 33 at 0.3 s per decade.

Spectra obtained from both phases were interpreted on a MassLib data system (Max Planck Institute, Germany), using IFF in-house libraries and commercial Wiley 7, NIST 98 and other libraries. The identification of flavor components was confirmed by interpretation of MS data and by relative GC retention indices based on a calibration with ethyl esters.

Results and Discussion

Figure 1 presents mass spectral TIC profile of the liquid/liquid extract of tequila. A total of 115 compounds were detected in the extract. Major peaks were early eluting alcohols, e.g. ethanol, isoamyl alcohol, 2-methylbutyl alcohol, propanol, and isobutanol. Ethyl esters such as ethyl decanoate and ethyl dodecanoate were also present in significant amounts. The presence of ethanol and other alcohols in tequila leads to the formation of esters and acetals. As seen in Table I, esters were the major class of compounds, comprising one third of the total number of compounds.

The Twister TIC in Figure 2 shows a wide spread of components. Twister analysis yielded 236 components, most of which were esters (Table I). Due to the low affinity of Twister for the early eluting alcohols, they were not as abundant in Twister analysis as they were in liquid/liquid extract. Compounds highest in concentration in Twister analysis were ethyl esters, namely, ethyl decanoate, ethyl dodecanoate, ethyl octanoate, ethyl 9-hexadecenoate, and ethyl hexadecanoate.

Table I. Number of Components Detected in Tequila with Each Sampling Technique

Compound Class	L/L Extract	Twister
Hydrocarbons	11	46
Alcohols	23	22
Aldehydes	7	16
Ketones	5	15
Acids	10	4
Esters	38	80
Lactones	2	0
Ethers	1	8
Furans	2	9
Acetals	7	19
Nitrogen compounds	2	0
Sulfur compounds	0	3
Sulfur and nitrogen compounds	1	0
Unknowns	6	14

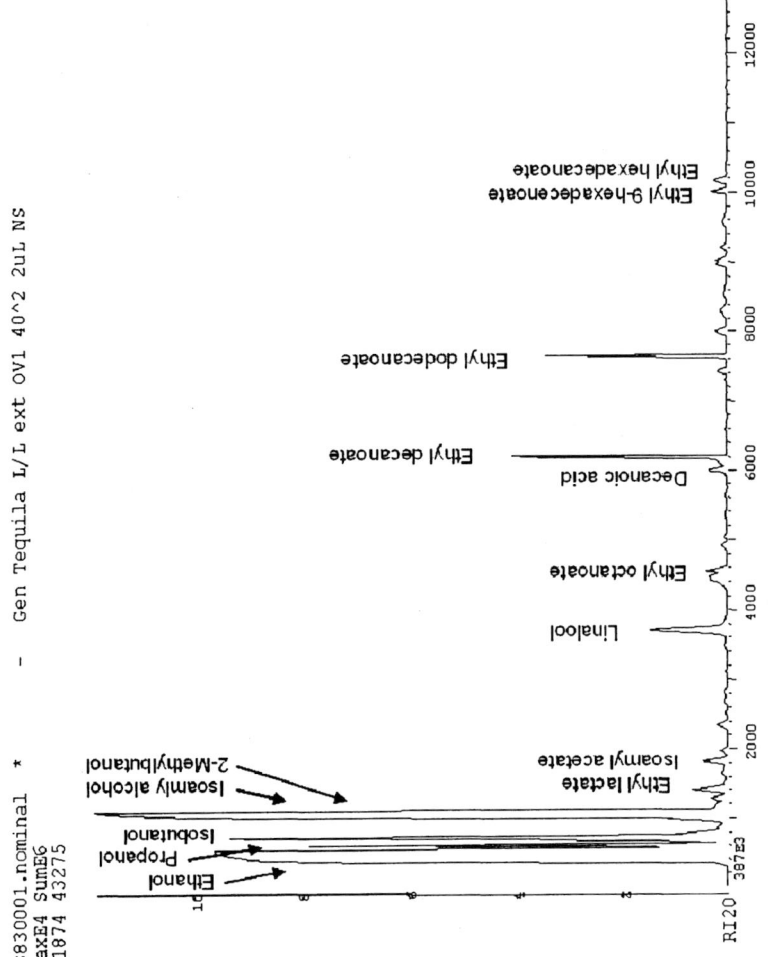

Figure 1. Mass spectral TIC profile of the liquid/liquid extract of tequila.

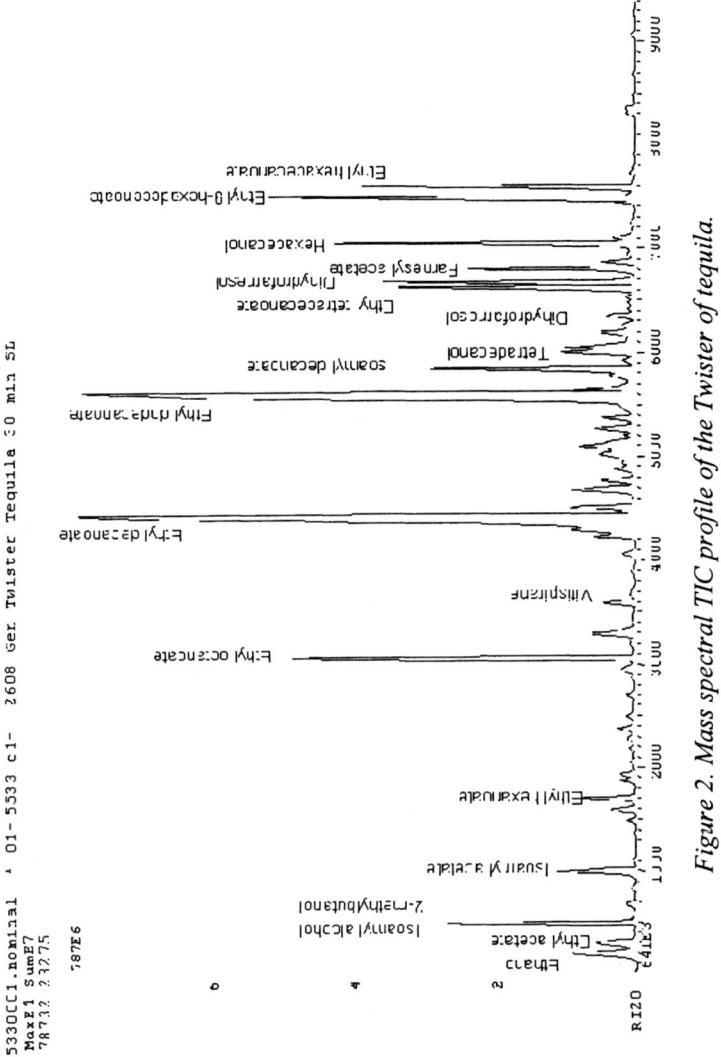

Figure 2. Mass spectral TIC profile of the Twister of tequila.

The orange peel oil contained 111 compounds. As seen from Figure 3, the major peak in the chromatogram was limonene. Valencene, myrcene, and α-pinene were also present in significant amounts. Terpene hydrocarbons were the predominant class of compounds; they comprised half of the total number of compounds (Table II) and were followed by aldehydes, alcohols, esters, ketones, and acids.

Freshly squeezed lime juice was analyzed by solvent extraction and headspace. Figure 4 shows the TIC profile of the solvent extract of lime juice. The major compound in the extract was limonene. Linalool and *cis*-3-hexenol were also present in significant amounts.

As seen from Figure 5, headspace analysis captured the most volatile compounds in the juice. Limonene was the major peak and it was followed by other terpene compounds, namely α-pinene, carvone, linalool, *cis-p*-mentha-2,8-dien-1-ol, and *trans-p*-mentha-2,8-dien-1-ol. Esters such as ethyl acetate, ethyl isobutyrate, and isoamyl acetate were also present at high concentrations.

Table III shows compounds detected in lime juice. A total of 104 compounds were detected in the solvent extract and 149 compounds were found in the headspace. Terpene hydrocarbons were predominant class in both analyses, comprising half of the total number of compounds. The other compound classes found in lime juice were alcohols, aldehydes, esters, ketones, ethers, acids (solvent extract and headspace), and furans (solvent extract).

Table IV lists compounds that were found in tequila, lime juice, and orange peel oil. Each analysis yielded a complex flavor profile. Esters were the predominant compound class in tequila, while lime juice and orange peel oil had more terpene hydrocarbons than any other class. All three ingredients contained large number of alcohols, aldehydes, and esters.

Table II. Compounds Detected in Orange Peel Oil

Compound Class	Peel Oil
Terpene hydrocarbons	53
Alcohols	15
Aldehydes	24
Ketones	3
Acids	1
Esters	14
Unknowns	1

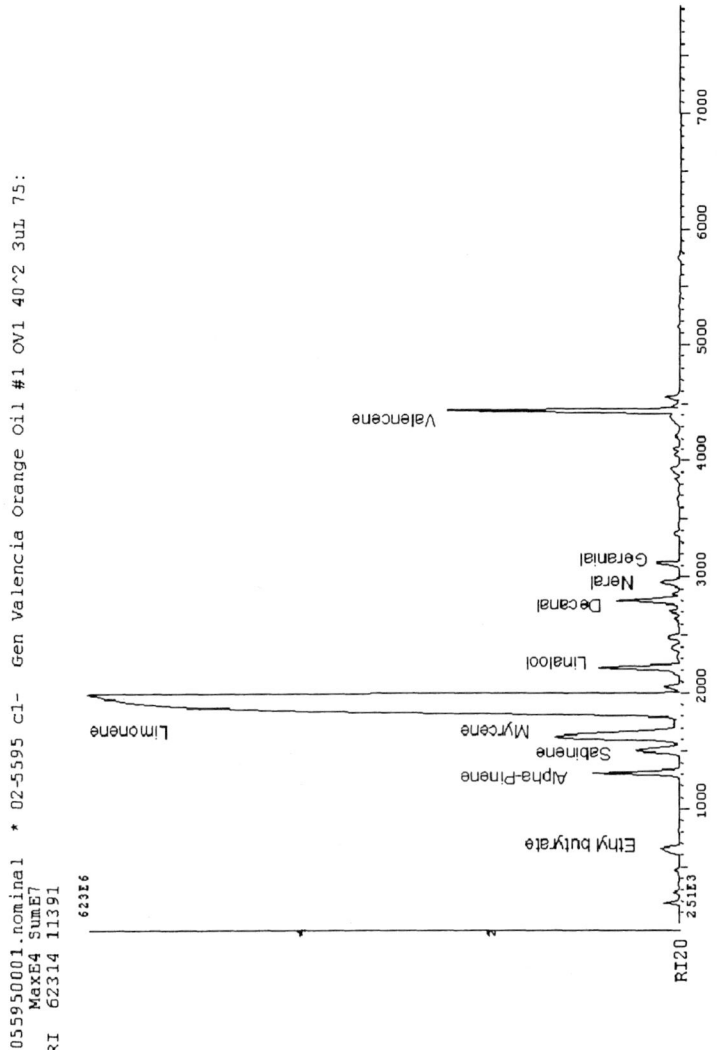

Figure 3. Mass spectral TIC profile of the Valencia orange peel oil.

188

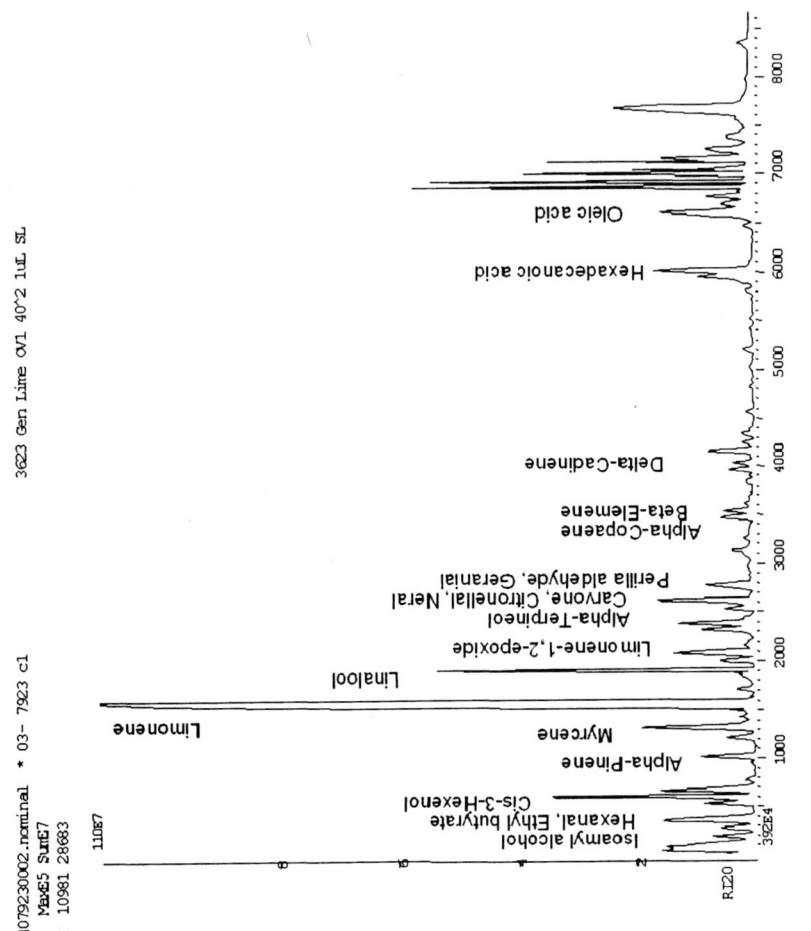

Figure 4. Mass spectral TIC profile of the liquid/liquid extract of lime juice.

189

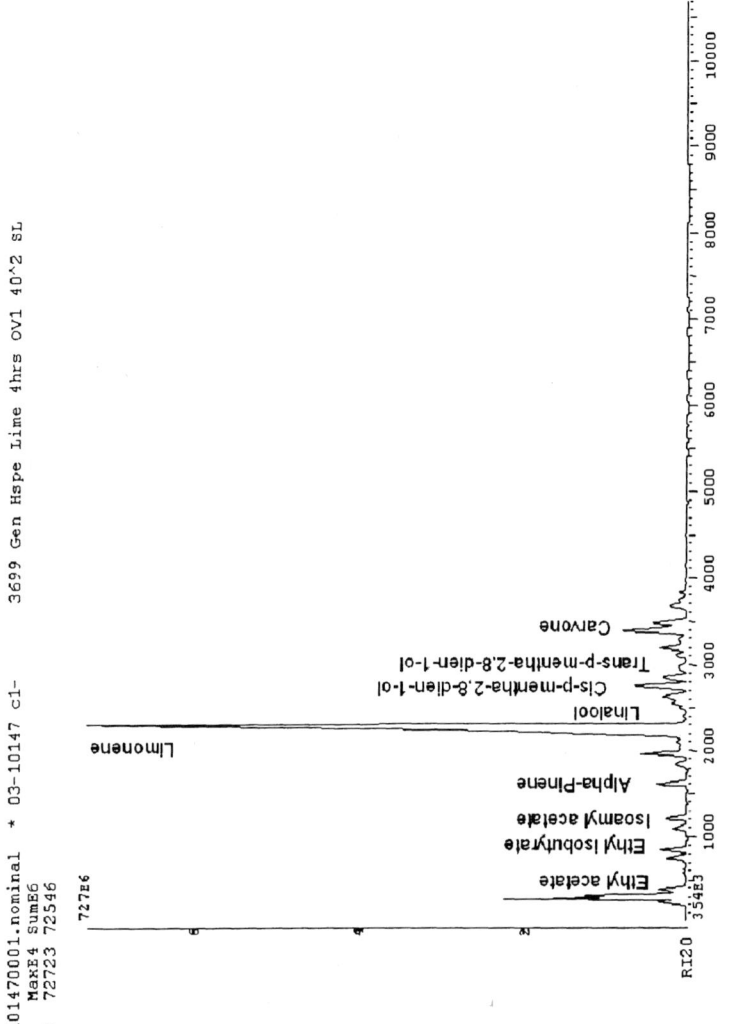

Figure 5. Mass spectral TIC profile of the headspace of lime juice.

Table III. Compounds Detected in Lime Juice

Compound Class	L/L Extract	Headspace
Terpene hydrocarbons	48	73
Alcohols	16	12
Aldehydes	11	26
Ketones	11	9
Acids	1	1
Esters	10	18
Ethers	4	1
Furans	3	0

Table IV. Compounds Identified in Tequila, Lime Juice, and Orange Peel Oil

Compound Class	Tequila L/L Extract	Lime Juice L/L Extract	Orange Peel Oil
Hydrocarbons	11	48	53
Alcohols	23	16	15
Aldehydes	7	11	24
Ketones	5	11	3
Acids	10	1	1
Esters	38	10	14
Lactones	2	0	0
Ethers	1	4	0
Furans	2	3	0
Acetals	7	0	0
Nitrogen compounds	2	0	0
Sulfur compounds	0	0	0
Sulfur and nitrogen compounds	1	0	0
Unknowns	6	0	1

Conclusion

The flavor of the classic Margarita cocktail is a complex one, comprised of several hundred compounds. Tequila has the most diverse flavor profile of compound classes, but contributes to the flavor mostly with esters and alcohols. Lime juice and orange peel oil (which is part of the orange liqueur) contribute mostly with terpenes.

Once the ingredients are mixed in the glass, new chemical reactions can occur. The presence of ethanol and other alcohols in tequila will lead to ester and acetal formation with lime juice and orange oil compounds.

References

1. Benn, S.M.; Peppard, T.L., *J. Agric. Food Chem.* **1996**, *44*, 557-566.
2. Lopez, M.G. In *Flavor and Chemistry of Ethnic Foods;* Shahidi, F.; Ho, C-T., Eds.; Kluwer Academic/Plenum Publishers: New York, 1997; pp 211-217.
3. Uhal, A.E.; Chisholm, M.G. 223rd ACS National Meeting, 2002, Abstract CHED730.
4. Kirbaslar, F.G.; Kirbaslar, S. I. *J. Essent. Oil Res.* **2003**, *15*, 6-9.
5. Stuart, G.; Lopes, D.; De Olivieira, J. *Perfumer & Flavorist* **2001**, *26*, 8-15.
6. Da Costa, N.C.; Eri, S. In *Chemistry and Technology of Flavors and Fragrances*; Rowe, D.J., Ed.; Blackwell Publishing: Oxford, UK, 2004; pp 12-31.
7. Parliament, T., In *Techniques for Analyzing Food Aroma*; Marsilli, R., Ed., Marcel Dekker, Inc.: New York, 1997; p 1-27.
8. Buttery, R.G.; Ling, L.C. In *Biotechnology for Improved Foods and Flavors;* Takeoka, G.R.; Teranishi, R.; Williams, J.; Kobayashi, A., Eds.; ACS Symposium Series 637; American Chemical Society: Washington D.C., 1996, pp 240-248.
9. Bicchi, C.; Iori, C.; Rubiolo, P.; Sandra, P. *J. Agric. Food Chem.* **2002**, *50*, 449-459.
10. Sandra, P.; David, F.; Vercammen, J. In *Advances in Flavours and Fragrances: from the Sensation to the Synthesis*; Karl A.D., Ed.; The Royal Society of Chemistry: London, 1995; pp 27-39.
11. Wampler, T.P. In *Techniques for Analyzing Food Aroma*; Marsili, R., Ed.; Marcel Dekker, Inc.: New York, 1997; pp 27-59.
12. Bicchi, C.; Cordero, C.; Iori, C.; Rubiolo, P. *J. High Res. Chrom.* **2000**, *23(9)*, 539-546.

Indexes

Author Index

Subject Index